PENNSYLVANIA COLLEGE OF TECHNOLOGY LIBRARY

5 0608 0 142607 8

DATE DUE

FEB 02 2012			

Env   rvival

Demco, Inc. 38-293

D1311699

ENVIRONMENT AND DEVELOPMENT

A fundamental element of sustainable development is environmental sustainability. Hence, this series was created in 2007 to cover current and emerging issues in order to promote debate and broaden the understanding of environmental challenges as integral to achieving equitable and sustained economic growth. The series will draw on analysis and practical experience from across the World Bank and from client countries. The manuscripts chosen for publication will be central to the implementation of the World Bank's Environment Strategy, and relevant to the development community, policy-makers, and academia. Topics addressed in this series will include environmental health, natural resources management, strategic environmental assessment, policy instruments, and environmental institutions.

Also in this series:
International Trade and Climate Change: Economic, Legal, and Institutional
    Perspectives
Poverty and the Environment: Understanding Linkages at the Household Level
Strategic Environmental Assessment for Policies: An Instrument for Good Governance

Environmental Health
and
Child Survival

Epidemiology, Economics, Experiences

THE WORLD BANK
Washington, DC

JUL 30 2009

Madigan Library
Pennsylvania College
of Technology
One College Avenue
Williamsport, PA 17701-5799

© 2008 The International Bank for Reconstruction and
Development / The World Bank
1818 H Street, NW
Washington, DC 20433
Telephone 202-473-1000
Internet www.worldbank.org
E-mail feedback@worldbank.org

All rights reserved.

1 2 3 4 :: 11 10 09 08

This volume is a product of the staff of the International Bank for
Reconstruction and Development / The World Bank. The findings, inter-
pretations, and conclusions expressed in this volume do not necessarily
reflect the views of the Executive Directors of The World Bank or the
governments they represent.

The World Bank does not guarantee the accuracy of the data included
in this work. The boundaries, colors, denominations, and other information shown on any map in this work do not imply any judgement on the
part of The World Bank concerning the legal status of any territory or the
endorsement or acceptance of such boundaries.

RIGHTS AND PERMISSIONS

The material in this publication is copyrighted. Copying and/or transmit-
ting portions or all of this work without permission may be a violation of
applicable law. The International Bank for Reconstruction and
Development / The World Bank encourages dissemination of its work and
will normally grant permission to reproduce portions of the work promptly.

For permission to photocopy or reprint any part of this work, please
send a request with complete information to the Copyright Clearance Center
Inc., 222 Rosewood Drive, Danvers, MA 01923, USA; telephone: 978-750-
8400; fax: 978-750-4470; Internet: www.copyright.com.

All other queries on rights and licenses, including subsidiary rights,
should be addressed to the Office of the Publisher, The World Bank,
1818 H Street NW, Washington, DC 20433, USA; fax: 202-522-2422;
e-mail: pubrights@worldbank.org.

ISBN-13: 978-0-8213-7236-4
eISBN-13: 978-0-8213-7237-1
DOI: 10.1596/978-0-8213-7236-4

Library of Congress Cataloging-in-Publication Data
Environmental health and child survival : epidemiology, economics, expe-
riences.
       p. ; cm. — (Environment and development)
  Includes bibliographical references and index.
  ISBN 978-0-8213-7236-4
 1. Environmentally induced diseases in children--Developing countries. 2.
Malnutrition in children--Developing countries. I. World Bank. II. Series:
Environment and development (Washington, D.C.)
  [DNLM: 1. Child, Preschool. 2. Environmental Health. 3. Cost of Illness.
4. Developing Countries. 5. Disorders of Environmental Origin. 6.
Malnutrition. WA 30.5 E605 2008]
  RJ383.E583 2008
  618.92'98--dc22

                                                2008022136

Cover photo:
World Bank Photo Library

Cover design:
Auras Design, Silver Spring, Maryland

CONTENTS

Boxes

Figures

Tables

Acknowledgments

This book is a product of the Environmental Health Anchor Program in the Environment Department of the World Bank. The book was prepared by a team led by Anjali Acharya (Environmental Specialist, ENV/World Bank) and Mikko K. Paunio (Sr. Environmental Specialist, ENV/World Bank) under the guidance of Kulsum Ahmed (Lead Environmental Specialist and Team Leader, Environmental Health Anchor Program, ENV/World Bank) and Laura Tlaiye (Sector Manager, ENV/World Bank). The core writing team also included Maria Fernanda Garcia (Consultant, ENV/World Bank), Monica Das Gupta (Sr. Social Scientist, DECRG/World Bank), Peter Kolsky (Sr. Water Sanitation Specialist, ETWWA/World Bank), Bjorn Larsen (Consultant, ENV/World Bank), and Giovanni Ruta (Economist, ENV/World Bank).

Special thanks go to the peer reviewers for this study, who included Harold Alderman (Advisor, AFTHD/World Bank), Enis Baris (Sr. Health Specialist, MNSHD/World Bank), Sandy Cairncross (Professor of Environmental Health, London School of Hygiene and Tropical Medicine), and Maureen Cropper (Professor of Economics, University of Maryland). Sandy's continuous and passionate engagement on the content and tone of this book is highly appreciated, while Maureen's role in providing substantial inputs to the economic costing methodology developed for part of this report is especially recognized.

Additional comments, inputs and guidance are gratefully acknowledged from Douglas Barnes (Sr. Energy Specialist, ETWES/World Bank), Caroline van den Berg (Sr. Economist, ETWWA/World Bank), Jan Bojö (Lead Environmental Economist, ENV/World Bank), Sandra Cointreau (Solid Waste Management Adviser, FEU/World Bank), James Listorti (Consultant, FEU/World Bank), Richard Seifman (Consultant, AFTHV/World Bank), and Kate Tulenko (Public Health Specialist, WSP/World Bank). The team would also like to thank Maria Neira (Director), Jamie Bartram, Carlos Corvalán and Annette Prüss-Üstün, from the World Health Organization's Department of Public Health and Environment, for sharing data relating to their new estimates of burden of disease from water, sanitation, and hygiene.

The support of the Bank-Netherlands Partnership Program in the preparation of this book is also gratefully acknowledged.

Abbreviations and Acronyms

AF	attributable fraction
AIDS	acquired immune deficiency syndrome
ALRI	acute lower respiratory infection
ARI	acute respiratory infection
DALY	disability-adjusted life year
DDT	dichloro-diphenyl-trichloroethane
DHS	Demographic and Health Survey
GDP	gross domestic product
HAZ	height for age z-score
HIV	human immunodeficiency virus
IAP	indoor air pollution
IMCI	Integrated Management of Childhood Illness (strategy)
ITN	insecticide-treated net
IUGR	intrauterine growth restriction
LSMS	Living Standards Measurement Survey
MAL	malnutrition
MDG	Millennium Development Goals
MICS	Multiple Indicator Cluster Survey
NGO	nongovernmental organization
NISP	National Improved Stove Program (China)
RR	relative risk
SD	standard deviation
UNICEF	United Nations Children's Fund
WAZ	weight for age z-score
WHO	World Health Organization
WRM	water resource management
WSH	water, sanitation, and hygiene
WTP	willingness to pay

CHAPTER 1

Introduction

INTEREST IN ENVIRONMENTAL HEALTH has mounted in recent years, spurred by concern that the most vulnerable groups—including children under five years of age—are disproportionately exposed to and affected by health risks from environmental hazards (see box 1.1). More than 40 percent of the global burden of disease attributed to environmental factors falls on children below five years of age, who account for only about 10 percent of the world's population (WHO 2007b). In large, populous areas in South Asia and Sub-Saharan Africa, where environmental health problems are especially severe, malnutrition in young children is also rampant.

Malnutrition is an important contributor to child mortality. Today, in low-income countries, more than 147 million children under the age of five remain chronically undernourished or stunted, and more than 126 million are underweight (Svedberg 2006; World Bank 2006c). Children in the developing world continue to face an onslaught of disease and death from largely preventable factors. These children are especially susceptible to these environmental factors, which put them at risk of developing illness in early life. Acute respiratory infections annually kill an estimated 2 million children under the age of five; 800,000 of those deaths are from indoor air pollution (WHO 2007b). Diarrheal diseases claim the lives of nearly 2 million children every year; most of those deaths are attributed to contaminated water and inadequate sanitation and hygiene (WHO 2007b). Each year, approximately 300 million to 500 million malaria infections

1

BOX 1.1
What Is Environmental Health?

Environmental health is defined as those health outcomes that are a result of environmental risk factors. The World Health Organization has defined environmental health as "all the physical, chemical, and biological factors external to a person, and all the related factors impacting behaviours. It encompasses the assessment and control of those environmental factors that can potentially affect health. It is targeted towards preventing disease and creating health-supportive environments" (WHO 2008). This study incorporates only those environmental health issues that relate to children— primarily water, sanitation, and hygiene; indoor air pollution; and malaria. These problems cause the top three diseases that affect children in developing countries.

Sources: Breman, Alilio, and Mills 2004; Ezzati, Rodgers, and others 2004; WHO 2008.

---

lead to more than 1 million deaths, of which more than 75 percent occur in African children under five years of age.

Malnutrition and environmental infections are inextricably linked; however, over time, these links have been forgotten or neglected by policy-makers in their formulation of strategies aimed at child survival and development. Persistent malnutrition and rampant environmental health problems are contributing to the widespread failure among developing countries to meet several of their commitments toward the Millennium Development Goals (MDGs), including not only the goal to halve poverty and hunger (MDG 1), but also the potential to halve maternal and child mortality (MDGs 4 and 5), to achieve universal primary education (MDG 2), to promote gender equality (MDG 3), and to combat malaria and confront the HIV/AIDS pandemic (MDG 6) by 2015 (see table 1.1). Research indicates that globally under-five mortality has fallen from 100 per 1,000 live births in 1980 to 72 per 1,000 in 2005. It is expected that the under-five death rate for the world will fall by 37 percent from 1990 to 2015, substantially less than the MDG 4 target of a 67 percent decrease (Murray and others 2007). Environmental health can contribute to many of the MDGs, as is shown in table 1.1.

In many developing countries, programs to improve child health have focused on improved feeding practices, micronutrient supplementation, national immunization campaigns, and measures to strengthen health systems (such measures include improving the availability of drugs, ensuring better treatment of cases, and hiring more trained personnel). However, with continued exposure to contaminated water, inadequate sanitation, smoke and dust, and mosquitoes, children in developing countries are still falling sick, a problem that imposes a sustained and heavy burden on the health system. And with the recognition of the environment's

TABLE 1.1
Millennium Development Goals and Environmental Health

Millennium Development Goal	Environmental Health Determinants Relating to Child Health
1. To eradicate extreme poverty and hunger	• Expenses incurred for informal sector delivery of water and sanitation services, as well as costs of medical treatment, impose a burden on family budgets (including food budgets). Lack of adequate water and sanitation services leads to diarrhea. These problems affect children's nutritional status adversely and indirectly add to a vicious cycle of poverty. • In urban areas, time spent fetching or queuing for water limits earning capacity.
2. To achieve universal primary education	The environmental health burden has significant effects on school performance and attendance.
3. To promote gender equality and empower women	• Women disproportionately suffer from (a) exposure to smoke from use of biomass for cooking, (b) drudgery and inconvenience from poor access to water, and (c) privacy and dignity issues related to inadequate sanitation. • Time spent collecting water and firewood impinges on time to care for sick children or to seek livelihood opportunities.
4. To reduce child mortality	Leading causes of child mortality include diarrhea, acute respiratory infections, and malaria. Indoor air pollution adversely affects young children (exposure to smoke from biomass use). Sickness and deaths result from inadequate hygiene, water supply, and sanitation.
5. To improve maternal health	• Inadequate hygiene and lack of availability of clean water results in poor health outcomes related to delivery and birthing. • Malaria and helminths affect pregnant women and can lead to malnutrition of the fetus.
6. To combat HIV/AIDS, malaria, and other diseases	• HIV-infected children especially need clean environments. • Environmental conditions related to mosquito breeding (such as lack of irrigation, poor drainage, and stagnant water) point to the need for adequate water resource management practices.
7. To ensure environmental sustainability	• Access to water and sanitation is a goal in itself. • Slum dwellers (including children) face dismal living conditions, congested settlements, and poor access to environmental services.
8. To establish a global partnership for development	Multisectoral coordination on environmental health issues is lacking. Both horizontal and vertical links are needed.

Source: Compiled by World Bank team.

contribution to malnutrition, there is an urgent need to broaden the spectrum of interventions beyond the health sector.

Objectives

The World Bank (2006c) study titled *Repositioning Nutrition as Central to Development* placed nutrition as a central issue to achieving the MDGs and established that malnutrition is not only due to lack of food but also the result of environmental risk factors. This report complements that study by looking at environmental health issues that affect child health broadly, while also exploring the links through malnutrition. This report argues that environmental health interventions are preventive measures that are imperative to improve child survival with sustainable results in the long term. Preventive measures—such as improving environmental conditions—are effective in reducing a child's exposure to a disease agent and thereby averting infection (Murphy, Stanton, and Galbraith 1997).

The overall aim of this report is to provide information to decision-makers on the optimal design of policies to help reduce premature deaths and illness in children under five years of age. To protect the health, development, and well-being of young children, decision-makers must identify and reduce environmental risk factors by providing appropriate interventions that prevent and diminish exposures. This study is intended to advance the understanding of what those risk factors are, when and how to reduce children's exposure to them, and how to mitigate their consequent health impact. Accordingly, the study has the following objectives:

- To provide an improved understanding of the links between environmental health risks and malnutrition through a review of literature and research. Moreover, the study discusses the role of environmental health inputs in a child's survival and growth.
- To analyze new data for the environmental health burden of disease (at a subregional level) that relates to children under five. These data, which are from a World Health Organization (WHO) report (Fewtrell and others 2007), include the total effects of environmental risk factors on health outcomes (including those mediated through malnutrition). Using two country examples, the study calculates the associated economic costs (including the costs of cognitive and learning impacts and of future work productivity).
- To highlight—through illustrative examples—how environmental health interventions are being delivered in developing countries through a variety of health, infrastructure, and environmental programs. The study also discusses the institutional and governance implications of delivering such multisectoral interventions.

Audience

The main audience for this report is senior policy-makers (and their technical staffs) who work in the ministries of planning, finance, health, environment, rural development, and infrastructure in developing countries and who are involved in designing policies for and allocating resources to programs that contribute toward improving child health. The study will also be useful to state- and local-level governments, because the actual implementation of programs and initiatives on child health is at the level of communities and households. Furthermore, donors and other organizations financing child health improvement initiatives and projects will benefit from a discussion of how interventions addressing environmental risks are important complements to health sector programs such as micronutrient supplementation and vaccination campaigns. Finally, health, environment, and infrastructure specialists working in developing countries will also gain from understanding the importance of working on children's health from different angles in a harmonized, constructive, and collaborative way.

A Primer on Environmental Health

Environmental health relates to human activity or environmental factors that have an impact on socioeconomic and environmental conditions with the potential to reduce human disease, injury, and death, especially among vulnerable groups—mainly the poor, women, and children under five (Listorti and Doumani 2001; Lvovsky 2001). The top killers of children under five are acute respiratory infections (from indoor air pollution); diarrheal diseases (mostly from poor water, sanitation, and hygiene); and malaria (from inadequate environmental management and vector control). This report concentrates on three specific environmental risk factors that influence a child's health: (a) poor water, sanitation, and hygiene; (b) indoor air pollution; and (c) inadequate malaria vector control.

Poor Water and Sanitation Access

With 1.1 billion people lacking access to safe drinking water and 2.6 billion without adequate sanitation, the magnitude of the water and sanitation problem remains significant (WHO and UNICEF 2005). Each year contaminated water and poor sanitation contribute to 5.4 billion cases of diarrhea worldwide and 1.6 million deaths, mostly among children under the age of five (Hutton and Haller 2004). Intestinal worms—which thrive in poor sanitary conditions—infect close to 90 percent of children in the developing world and, depending on the severity of the infection, may lead to malnutrition, anemia, or retarded growth, which, in turn, leads to diminished school performance (see Hotez and others 2006; UNICEF 2006). About 6 million people are blind from trachoma, a disease caused by the lack of water combined with poor hygiene practices.

Indoor Air Pollution

Indoor air pollution—a much less publicized source of poor health—is responsible for more than 1.6 million deaths per year and for 2.7 percent of global burden of disease (Smith, Mehta, and Maeusezahl-Feuz 2004; WHO 2006). It is estimated that half of the world's population, mainly in developing countries, uses solid fuels (biomass and coal) for household cooking and space heating (Rehfuess, Mehta, and Prüss-Üstün 2006). Cooking and heating with such solid fuels on open fires or stoves without chimneys lead to indoor air pollution, which, in turn, results in respiratory infections. Exposure to these health-damaging pollutants is particularly high among women and children in developing countries, who spend the most time inside the household. As many as half of the deaths attributable to indoor use of solid fuel are of children under the age of five (Smith, Mehta, and Maeusezahl-Feuz 2004).

Malaria

Approximately 40 percent of the world's people—mostly those living in the world's poorest countries—are at risk from malaria. Every year, more than 500 million people become severely ill with malaria, with most cases and deaths found in Sub-Saharan Africa. However, Asia, Latin America, the Middle East, and parts of Europe are also affected. Pregnant women are especially at high risk of malaria. Nonimmune pregnant women risk both acute and severe clinical disease, resulting in fetal loss in up to 60 percent of such women and maternal deaths in more than 10 percent, including a 50 percent mortality rate for those with severe disease. Semi-immune pregnant women with malaria infection risk severe anemia and impaired fetal growth, even if they show no signs of acute clinical disease. An estimated 10,000 women and 200,000 infants die annually as a result of malaria infection during pregnancy (WHO 2007d).

A Primer on Malnutrition

Malnutrition remains an underlying cause of death in half of the 10.5 million deaths globally in children under five (Bryce and others 2005). In low-income countries, more than 147 million (or 27 percent) children under the age of five remain chronically undernourished or stunted, and more than 126 million (or 23 percent) are underweight. South Asia, where about one-fifth of the world population lives, still has both the highest rates and the largest numbers of malnourished children in the world. In Afghanistan, Bangladesh, India, and Pakistan, the prevalence rate varies from 38 to 51 percent and is only gradually declining, whereas in Sub-Saharan Africa, while the rate is lower at 26 percent, it is on the rise (Svedberg 2006; World Bank 2006c).

Although lack of food is obviously an important reason for malnutrition, recent reports and studies ever more consistently suggest that much of malnutrition is actually caused by bad sanitation and disease, especially in young children (WHO 2007e; World Bank 2006c). Thus, contrary to popular perception, in many countries where malnutrition is widespread, insufficient food production is often not the determining factor of malnutrition (Prüss-Üstün and Corvalán 2006; World Bank 2006c). A recent collective expert opinion stated that about 50 percent of the consequences of malnutrition are in fact caused by inadequate water and sanitation provisions and poor hygienic practices (Prüss-Üstün and Corvalán 2006), thus highlighting the need to mainstream environmental health into the development agenda.

Nutrition in early childhood—starting from the womb—is critical for child health and, consequently, for adult health. Maternal anemia in pregnant women—caused from a combination of malaria and hookworm infections—leads to malnourishment of the fetus, a condition called *intrauterine growth restriction* (IUGR). Babies suffer from low birth weight in developing countries mostly because of IUGR, whereas in developed countries, the condition is far more often attributable to preterm birth. Repeated infections—especially diarrhea and helminths—caused by poor environmental conditions lead to underweight (low weight for age) and stunted (low height for age) children. These growth-faltering effects, in turn, make individuals more predisposed to infections and even to chronic diseases later in life.

Commonly used indicators of malnutrition are underweight, stunting, and wasting. *Underweight* is measured as the child's weight for age relative to an international reference population. *Stunting* is measured as the child's height for age, and *wasting* is measured as the weight for height. Underweight is an indicator of chronic or acute malnutrition or a combination of both. Stunting is an indicator of chronic malnutrition, and wasting an indicator of acute malnutrition. How far a child's measure is from the mean of the reference population—measured in standard deviations (SDs) from the mean—determines the extent of malnutrition: mild (–1 SD to –2 SD), moderate (–2 SD to –3SD), or severe (greater than –3SD).

Childhood malnutrition is associated with increased susceptibility to disease and with poor mental development and learning ability. In the long term, those outcomes are a significant cost to countries (Alderman and others 2006). Although research and mainstream debate have revolved around how malnourished children are more susceptible to infectious diseases (including diarrhea and acute respiratory infections), the extent to which environmental risk factors contribute to malnutrition is not widely acknowledged.

Content and Organization

This report is organized into three main sections: the first looks at the *epidemiology* (science and research evidence), the second presents the *economics* (costs of the burden of disease and costs related to learning deficits and productivity losses), and the third describes the *experiences* of environmental health actions in developing countries. Each section strives to present the latest information and data and highlights the reasons environmental health is so critical in the context of child survival and development.

Epidemiology

Chapter 2 argues that improvements in environmental health are very important for child survival and development, especially considering its links through malnutrition. The epidemiological underpinnings of the infections-malnutrition cycle are important because repeated infections cause a decrease in dietary intake, producing, for example, malabsorption of nutrients, which in effect causes malnutrition, thereby making children weak in resisting disease and likely to fall sick again.

Until recently, the impact of diseases such as diarrhea and respiratory infections on malnutrition in children was relatively ignored. Over the past several decades, dozens of studies—many of them long-term cohort studies—have investigated the causal relationship between disease and malnutrition. These cohort studies have provided strong evidence of how almost all infections influence a child's nutritional status. A review of the studies was carried out for this report and served to provide further corroboration of the impacts of environmental infections on child growth, including through malnutrition. Evidence from several of the studies demonstrates how exposure to environmental health risks in early infancy leads to permanent growth faltering, lowered immunity, and increased morbidity and mortality.

Environmental health inputs—both at the household and the community levels—play a critical role in a child's survival and growth. In the life cycle of a child, environmental health interventions are critical, especially in the period from the womb to the age of about two years. This period is the so-called window of opportunity. Pregnant women in developing countries are often exposed to environmental risks such as malaria and hookworm infections, which contribute to poor fetal growth and result in babies with low birth weights. Smoky kitchens from use of biomass fuels have anecdotally revealed impacts on low birth weight and perinatal mortality. In early infancy, improper feeding practices and poor sanitation have a pernicious synergistic effect on the child's nutritional status. Many of these impacts on a child's growth have also been seen to result in cognition and learning impacts as well as chronic diseases later in life.

Current child survival strategies in developing countries mostly adopt a more treatment-oriented perspective, relying mainly on case management and focusing primarily on reducing mortality. Most of these strategies, while intended to increase the ability of the host to resist or reduce infection once exposure has occurred, do not attempt to reduce the exposure to environmental determinants of ill health.

Chapter 3 explores how appropriate environmental health actions can complement and supplement strategies that focus on child health by adding value to
health systems, by assisting in the adaptation of environmental management
programs, and by promoting adjustments to infrastructure strategies.

Economics

Chapter 4 provides key information and data relating to the burden of disease
from environmental factors and to the associated economic costs. Measuring the
burden of disease and subsequent economic costs from environmental health
risks is important in helping policy-makers better integrate environmental health
into economic development and, specifically, into their decisions relating to the
allocation of resources among various programs and activities to improve child
health. Building on previous estimates and taking into consideration the links
between environmental health, malnutrition, and disease, WHO recently revised
the burden of disease estimates taking into account malnutrition-mediated health
impacts associated with inadequate water and sanitation coverage and improper
hygienic practices (Fewtrell and others 2007).

The new WHO estimates reveal that the environmental health burden in children under five years is substantially higher when all links through malnutrition
are incorporated. This finding is especially apparent in subregions such as Sub-
Saharan Africa and South Asia, where malnutrition and poor environmental
conditions coexist. In Sub-Saharan Africa, despite much poorer living standards,
fewer babies are born with low birth weight than in South Asia. This enigma may
in part be explained by the poor survival rate of both fetuses and children in Sub-
Saharan Africa as a result of unhealthy environmental conditions. Furthermore,
even when conservatively estimated, a multiplier effect exists for environmental
health interventions: investments addressing environmental risks (such as lack of
water and sanitation) not only reduce diarrheal mortality but also reduce mortality
from malnutrition-related diseases and its consequences on education attainment.

Using case studies from Ghana and Pakistan, chapter 5 translates the burden
into economic costs at a country level. In doing so, it updates earlier estimates by
providing measures of the total effects of environmental risks, including those
through malnutrition. Also, the report for the first time attempts to estimate the
longer-term impacts of these environmental health risks on cognition and learning
and on future work productivity. These revised estimates show that when
malnutrition-mediated health effects attributed to environmental health risks
are included, the total costs for Ghana and Pakistan range from 4 to 6 percent of
a country's gross domestic product (GDP) (see table 1.2). These costs are at least
40 percent higher than when malnutrition-mediated effects are not included.

In the longer term, malnutrition (which is partly attributed to environment-
related infections) is found to affect a child's cognitive function, school enrollment,

TABLE 1.2
Annual Cost of Direct and Indirect Impact of Environmental Risk Factors in 2005

	Ghana				Pakistan			
	Annual Deaths	Cost (₵ million)	Cost (US$ million)	Cost (% of GDP)	Annual Deaths	Cost (PRs billion)	Cost (US$ million)	Cost (% of GDP)
Estimation Excluding Malnutrition-Mediated Effects								
Mortality effects	24,712	371	412	3.84	131,611	195	3,250	2.90
Estimation Including Malnutrition-Mediated Effects								
Mortality effects	35,702	537	595	5.55	187,429	278	4,633	4.13
Education effects		367	407	3.79		317	5,281	4.71
Total effects		904	1,002	9.34		595	9,914	8.84

Source: Compiled by World Bank team.
Note: ₵ = Ghanaian new cedi.

grade repetition, school dropout rate, grade attainment, and future income-earning potential. For Ghana and Pakistan, the annual cost of stunting attributable to early childhood diarrheal infections is estimated to be 4 to 5 percent of the country's GDP.

To estimate malnutrition-mediated costs, these analyses often rely on parameters from global and regional studies when corresponding country-level data are unavailable. Overall, wherever assumptions are required, the parameters have been conservatively chosen. Thus, when all effects through malnutrition are considered (including education costs), the total estimated annual costs may be as high as 9 percent of a country's GDP (see table 1.2). This social and economic burden is not trivial. It highlights the urgent need for policy-makers to position environmental health at the center of all child survival strategies.

Experiences

Chapter 6 begins with a historical review of environmental health, outlining the trends in the evolution of environmental health functions in developed countries and highlighting how circumstances have led to the unfortunate neglect of environmental health in the development agenda. Environmental health actions are the earliest public health activities on record. Lessons from history have shown the enormous benefits of multisectoral environmental health actions, with today's developed countries having undergone an evolution in environmental health functions. However, both institutionally and conceptually, environmental health has fallen through the cracks in the development agenda in the world's poorest countries.

On the one hand, there has been a growing environmental movement, with the creation of ministries of environment and accompanying policies and regulations. On the other hand, health ministries and state health departments have been engaged in scaling up vertical health sector programs, which focus mainly on treatment. This artificial separation of environment from traditional public health functions allows only limited multisectoral action that is needed to tackle environmental risks facing children under five.

Chapter 6 then presents illustrative examples of how different developing countries have incorporated environmental health activities within other health, nutrition, and infrastructure programs. Developing countries vary considerably in terms of institutional capacity, political will, and socioeconomic development. Environmental health interventions therefore need to be customized to the specific enabling environment in a developing country. Recognizing those differences, rather than providing specific recommendations, the chapter presents some illustrative examples of ways in which some developing countries are beginning to mainstream environmental health components and objectives within existing child survival programs, nutrition initiatives, and infrastructure projects (water and sanitation or rural energy projects).

Some common elements for successful environmental health actions in developing countries have included garnering high-level political commitment, involving and empowering communities, allocating responsibilities and resources at the local level, and finding a balance between private and public sector roles. Furthermore, successful environmental health governance requires strong institutional underpinnings, with clearly articulated roles at all levels of administration within a country. The study provides a discussion of the roles that national and local governments, as well as the international community, can play in delivering and managing environmental health interventions.

Now is a critical time for this agenda to take the forefront in developing countries, with governments, donors, and civil society beginning to strengthen measures to address environmental health, especially in the context of child survival. Chapter 7 highlights several key conclusions of this report:

- Diseases from environmental risk factors—diarrheal diseases, acute respiratory infections, and malaria—remain the top killers of children under five in developing countries. Research evidence of the cycle of disease (infections) and malnutrition implies a larger role for environmental health, because it also indirectly contributes to other diseases by weakening the child's immunity (through malnutrition). Environmental health actions also improve the effectiveness of other child health strategies.

- Specific subregions of the world—such as Sub-Saharan Africa and South Asia, where poor environmental conditions and high malnutrition prevalence coexist—should be especially targeted to fund and implement environmental health interventions. The multiplier effect of such interventions points to the potential of their significant health externalities. At a country level, the burden of disease associated with environmental risk factors is as high as 9 percent of a country's GDP—significant enough for policy-makers to consider environmental health programs in resource allocation decisions.

- Developing countries can learn from the experience of developed countries in addressing public health risks. More than 150 years ago, today's developed countries made deliberate attempts to improve environmental health conditions by specifically addressing sanitation and air pollution issues with relatively cost-effective interventions. Recent experience from, for example, the Ethiopian sanitation revolution has shown that improvements in rural settings in developing countries can also be achieved with modest fiscal inputs (WSP 2007), but such improvements need to be backed by political support and community involvement.

The child health agenda remains unfinished in the developing world, with millions of children continuing to fall sick and die from preventable environmental health causes. Although considerable progress has been made, the potential of environmental health actions to complement existing health, infrastructure, and environment management strategies remains largely untapped in the developing world.

In many ways, this report represents a first step toward providing policy-makers with the epidemiological, economic, and experiential evidence to incorporate environmental health into the child survival agenda. However, additional research and studies will help donors and governments in developing countries choose appropriate environmental health interventions. Such research efforts should include the following:

- Further research on environmental health impacts during pregnancy, on additional disease transmission pathways, and on better relative risk estimates will help improve disease burden and costing estimates.
- At a country level, cost-effectiveness and cost-benefit analyses are important follow-up exercises that will help guide decision-makers to prioritize among the various available interventions.
- A more in-depth country-level institutional analysis is required of the coordination mechanisms between ministries and of the ways mandates and budgets are assigned. Such analysis would help guide the roles and responsibilities of different agencies for better environmental health governance.

In the longer term, environmental health concerns are expected to grow. As the world's climate changes, diseases such as diarrhea and malaria, among other important health burdens that are the result of environmental risk factors, are likely to worsen, particularly for the poor in developing countries (Campbell-Lendrum, Corvalán, and Neira 2007; IPCC 2007). Thus, scaling up preventive environmental health interventions to reduce the current burden of disease is a prudent investment (Campbell-Lendrum, Corvalán, and Neira 2007).

Given the multisectoral nature of environmental health issues, the advocacy and regulatory roles of the health sector and the supporting roles of other sectors (such as the environmental, infrastructure, agricultural, and education sectors) in promoting and delivering environmental health actions need to be revitalized. Ultimately, good environmental health governance will require policy-makers to develop signaling mechanisms to identify environmental risks, to translate these signals into appropriate interventions, to adjust their policies to better address environmental health outcomes, and to set up institutional mechanisms to successfully implement interventions.

A concerted and continuous effort is needed on behalf of both developed and developing countries to ensure that environmental health is placed high on the development agenda, and corresponding interventions must be financed and undertaken to improve children's survival and development potential.

PART I

Epidemiology

CHAPTER 2

Environmental Health, Malnutrition, and Child Health

MALNUTRITION, POOR ENVIRONMENTAL CONDITIONS, and infectious diseases are highly associated geographically and take their heaviest tolls on children under five years of age in Sub-Saharan Africa, South Asia, and certain countries in the Eastern Mediterranean region (Ezzati, Vander Hoorn, and others 2004; Ezzati, Rodgers, and others 2004). Malnutrition is an underlying cause of child mortality that contributes to between 34.5 percent and 52.5 percent of the 10.5 million deaths globally in children under five (Caulfield and others 2006; Fishman and others 2004). In addition, childhood malnutrition is associated with disease, poor mental development, and reduced learning ability (Alderman and others 2006). Because research and mainstream debate have revolved around how malnourished children are more susceptible to infectious diseases—including diarrhea and acute respiratory infection (ARI)—the links between environmental risk factors and malnutrition are less acknowledged.

This chapter revisits the links between malnutrition and environment-related infections and seeks to demonstrate the importance of environmental health in child survival and growth. An overview of past and recent research that shows the importance of repeated disease episodes (such as diarrhea and malaria) in the development of malnutrition is presented (for a detailed analysis see appendix A).

Environmental Factors, Exposure, and Transmission Pathways

Environmental health focuses on disease transmission routes rather than on how people are treated when they are sick. The identification of transmission routes, rather than the diseases themselves, is the important conceptual framework, and because diseases can be transmitted by more than one route, environmental health interventions often make more sense at a community level than at the level of individuals (Yacoob and Kelly 1999). Table 2.1 gives examples of the different transmission routes that various water-related diseases can take. Such transmission routes have largely been blocked in the developed world. In developing countries, the poor continue to be exposed to many transmission routes at one time.

Understanding how different transmission routes affect disease outcomes—especially for diarrhea—is important because even when an intervention may aim at blocking one transmission route, the effect on the disease may be limited because the population is still exposed through another transmission route (see figure 2.1). This concept is known as *residual transmission* (Briscoe 1987; Cairncross 1987; Eisenberg, Scott, and Porco 2007). As Cairncross and Valdmanis (2006: 775) point out, "practically all potentially waterborne infections that are transmitted by the feco-oral route can potentially be transmitted by other means (contamination of fingers, food, fomites, field crops, other fluids, flies, and so on) all of which are water-washed routes."

The pervasive nature of fecal pollution in developing countries (see, for example, Kimani-Murage and Ngindu 2007) makes effective prevention of disease by blocking just one transmission route difficult (Eisenberg, Scott, and Porco

TABLE 2.1
Water-Related Transmission Routes and Disease Outcome

Transmission Route	Description	Disease Group	Examples
Waterborne	Pathogen is ingested in drinking water	Feco-oral infections	Diarrhea, dysentery, typhoid fever
Water-washed	Transmission by inadequate water for hygiene conditions and practices	Most feco-oral, oro-oral, acute respiratory, skin and eye infections	Diarrhea, dysentery, typhoid fever, acute lower respiratory infections, scabies, and trachoma
Water-based	Transmission by means of aquatic invertebrate host	Water-based infections	Schistosomiasis, guinea worm
Water-related insect vector	Transmission by insect vector that breeds in or near water	Water-related insect vector infections	Dengue, malaria, trypanosomiasis

Source: Adapted from Cairncross and Valdmanis 2006, table 41.1.

FIGURE 2.1
The F-Diagram: Transmission Routes for Infection

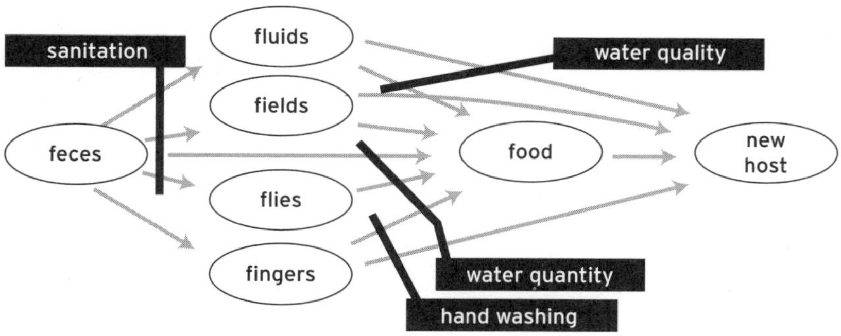

Source: Hunt 2006.

2007). The provision of clean water (one transmission route) has often produced less-than-anticipated outcomes because water may be scarce or hygiene practices poor (thus exposure remains through another transmission route). The emphasis on drinking water possibly occurs because those living in affluent conditions (with substantial water quantity[1] and proper sanitation in their homes) often ignore the other water-washed transmission routes for diseases that poorer households (with inadequate water for proper hygiene practices) face. This idea also has its roots in the historical drama of single-source epidemics rather than in the long-term tragedy of endemic diarrhea.

Vicious Cycle of Infections and Malnutrition

Infections and malnutrition operate in a vicious cycle to affect child health. Though the effect of malnutrition on disease is generally recognized, the role of infections in the worsening of nutritional status has been relatively neglected.

Effect of Malnutrition on Disease

Poor nutritional status, especially in infants and young children, makes infections worse and often more frequent. Data from a number of studies reviewed by Scrimshaw, Taylor, and Gordon (1968) provide evidence that moderate and severe undernutrition increases the seriousness of infections such as diarrhea and acute lower respiratory infection. Increased mortality is an effect of malnutrition, which makes individuals susceptible to infectious disease; when illness occurs, it is more severe and prolonged and carries an increased risk of death (Scrimshaw, Taylor, and Gordon 1968). As predicted in 1968, malnutrition was convincingly established as a *potentiator of mortality* in young children, with the risk of death from all infections increasing exponentially with decreasing nutritional status (Caulfield and others 2004; Fishman and others 2004; Pelletier 1994; Pelletier and others 1994).

Malnutrition can increase a child's susceptibility to infection by negatively affecting the barrier protection afforded by the skin and mucous membranes and by inducing alterations that reduce the child's immunity (Brown 2003) (see figure 2.2). For example, in a malnourished child, diarrhea can quickly result in life-threatening dehydration caused by loss of water and minerals (Thapar and Sanderson 2004). Malnutrition also increases the duration of many infections: the more severe the level of malnutrition, the longer the illness lasts, and the longer the child takes to recover (Thapar and Sanderson 2004).

Effect of Infections on Malnutrition

Up until the middle of the 20th century, nutrition textbooks hardly ever mentioned the role of infections in the worsening of nutritional status, which, in turn, reduces growth in children (Keusch 2003; Scrimshaw 2003). Even though, historically, vitamin deficiencies were known to be aggravated by infections, the effect of diseases such as diarrhea and respiratory infections on malnutrition in children was not recognized, and poor diets were considered to be predominantly responsible for poor growth in children (Scrimshaw 2003).

Over the past several decades, dozens of studies—many of them long-term cohort studies—have investigated the causal relationship between disease and malnutrition and have provided conclusive evidence of how almost all infections influence a child's nutritional status (see appendix A). Table 2.2 shows how infections adversely affect nutritional status in young children through reductions in food intake caused by loss of appetite as well as changes in intestinal absorption, changes in metabolism, and excretion of specific nutrients (Scrimshaw, Taylor,

FIGURE 2.2
Relationship between Nutrition and Infection

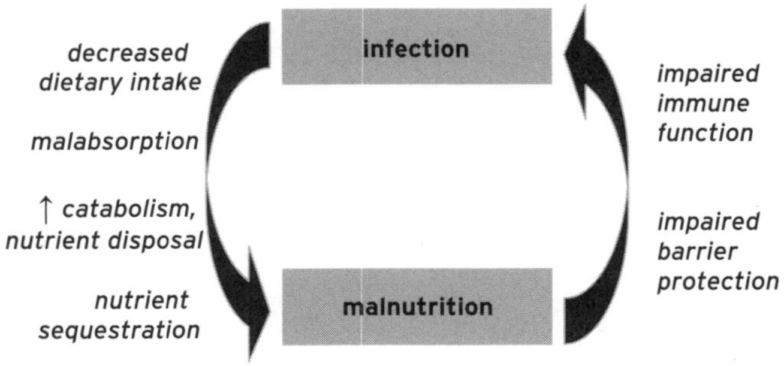

Source: Brown 2003.

TABLE 2.2
Impact of Infection on Nutritional Status

Infection	Effect	Nutritional Impact
ARI, diarrhea, measles (especially)	Anorexia	Decrease in food intake
Acute and chronic diarrhea (from bacteria, viruses, or intestinal helminths)	Malabsorption	Decrease in food intake
Diarrhea	Effusion of serous fluid to gut; traditionally "treated" by food withdrawal	Direct nutrient loss, mostly to gut or urine
Hookworms	Intestinal bleeding	
HIV infection or inflammation (especially in smokers)	Increased nutrient requirements	Increase in resting energy expenditure
Many infections	Several mechanisms among infected individuals leading to reduced serum retinol, iron, and zinc concentrations (relevance remains uncertain)	Diminished transport of micronutrients (vitamin A, iron, and zinc)

Source: Based on Stephensen 1999.
Note: For all the diseases, the stress, fever, or tissue damage caused by disease increases the food requirement while also reducing the intake and absorption of nutrients, which leads to anorexia.

and Gordon 1968; Stephensen 1999). These nutrient losses have implications for tissue synthesis and growth in young children—and lead to growth faltering (Brown 2003). The effect of infections on the nutritional status of young children appears to be directly proportional to the severity of the infection (Powanda and Beisel 2003), which means that children with more serious infections, such as dysentery, measles, or pneumonia, are more likely to become stunted than those with acute diarrhea (see appendix A).

Numerous cohort studies have researched the impact of infections (mostly diarrheal disease) on weight and linear growth (see box 2.1). A recent Peruvian study found that children ill with diarrhea 10 percent of the time during the first 24 months of life were 1.5 centimeters shorter than children who never had diarrhea (Checkley and others 2003). Another study from Brazil found that, on average, 9.1 diarrheal episodes before two years of age were associated with a 3.6 centimeter growth shortfall at age seven years (Moore and others 2001). Similarly, Moore and colleagues (2001) found that early childhood helminthiasis (infections caused by parasitic worms such as hookworms) led to a further 4.6 centimeter growth reduction by age seven. Because early childhood growth faltering is known to predict height in adulthood (Martorell 1995), the effects of infections on linear growth are considered irreversible (Checkley and others 2003). In general, effects on weight have been easier to demonstrate than effects on linear growth, even in shorter follow-up studies (Stephensen 1999).

BOX 2.1
Impact of Diarrhea on Child Malnutrition: Evidence from Research

Among the various infections, diarrhea is one of the most prevalent in developing countries and is responsible for a high proportion of sickness and death in children under five years of age (Scrimshaw 2003). Because of its high occurrence and its involvement with the malabsorption of nutrients, diarrhea has been a key issue in child malnutrition (Mata 1992; Scrimshaw 2003).

The effects of different types of malnutrition on diarrheal illness have been studied over the past several decades (Guerrant and others 1992) (see accompanying figure). Several studies in developing countries support the argument that an increased risk of diarrheal mortality is associated with low weight for age; however, the effect on diarrheal incidence has been less clear. Evidence now supports the idea that incidence is increased (Fishman and others 2004).

Effects of Diarrhea and Malnutrition

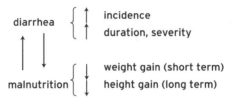

Source: Guerrant and others 1992.

Scrimshaw, Taylor, and Gordon (1959, 1968) summarized the evidence on how infections have a deleterious, although often ignored, effect on nutritional status. By 1968, the metabolic consequences of infections were already well established (Powanda and Beisel 2003). A major problem in the review was that long-term human follow-up information was lacking (Scrimshaw 2003). After 1968, about 38 cohort follow-up studies have produced information on this topic.

Practically all the cohort studies (see appendix A) favor the idea that not only diarrheal disease but also other infections (for example, helminths, measles, and acute lower respiratory infections) cause growth to falter. This view was strengthened because investigators could show larger effects on nutritional status with increasing frequency and severity of infections. Some inconsistencies and weak observed effects can be explained by the mitigating effects of breastfeeding and other factors that effectively mask effects of infections on growth faltering in these studies (see appendix A). Experimental clinical trial information on deworming of

(continued)

children further supports a causal inference between infection and poor growth (Taylor-Robinson, Jones, and Garner 2007).

The debate in the 1990s focused on whether these effects could be reversed, although even those who suggested that subsequent catch-up growth corrects much of the growth faltering recognize the failure in children with persistent diarrhea (that is, those who have diarrhea more than 10 percent of the time) (Bairagi and others 1987; Briend and others 1989; Moy and others 1994). Several studies during the past decade have shown that if infection burden begins before six months of age, lagging linear growth effects that are observed are likely to be irreversible (Adair and others 1993; Brush, Harrison, and Waterlow 1997; Checkley and others 2003; Moore and others 2001), which diminishes the possibility that malnourishment, in the first place, made the host more susceptible and that observed effects would result from reverse causality rather than true causality (see appendix A). A WHO collective expert opinion (Prüss-Üstün and Corvalán 2006) states that 50 percent of consequences of maternal and childhood underweight is attributable to lack of water, sanitation, and hygiene.

The idea that infections affect a child's nutritional status has faced some skepticism. Even while skeptics have agreed that, in the short term, the adverse metabolic effects from infections do lead to growth faltering in children, they have argued that subsequent "catch-up" growth fully compensates for these adverse effects of infection in the majority of children (Bairagi and others 1987; Briend and others 1989; Moy and others 1994; see also appendix A). Part of the reason for this skepticism and controversy about the impact of infections on growth faltering in children lay in the paucity of direct long-term human observations indicating the irreversibility of these effects (Scrimshaw 2003). Compelling evidence of this irreversibility was not gained until early 2000 (see appendix A).

Environmental Role in Early Childhood Health

Environmental health inputs—at both the household and the community levels—play a critical role in a child's survival and growth (see figure 2.3 and appendix A for further detail). In the life cycle of a child, from the womb to the age of about two years, environmental health interventions—such as access to water and sanitation, proper hygiene practices, proper vector control, and the use of cleaner fuels for cooking and heating—are especially critical for preventing growth faltering in the fetus and infant, which has consequences for a child's subsequent health. These impacts on a child's growth have also been seen to result in cognition and learning problems as well as chronic diseases later in life, an issue discussed in more detail later in this chapter.

FIGURE 2.3
Environmental Health Inputs and Health Outcomes in the Child's Life Cycle

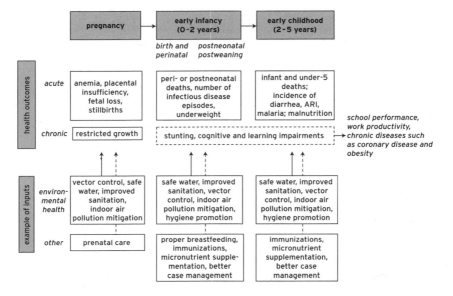

Source: Adapted from World Bank 2008a.

The following sections discuss the specific roles of environmental health actions in the life cycle of a child from pregnancy and protection of the fetus; to early infancy, when growth-faltering effects are irreversible; to early childhood, when both household- and community-level actions can make a difference. Finally, the longer-term effects, specifically in terms of cognition and learning, are discussed.

Pregnancy: Protecting the Fetus

Nutrition plays a crucial role in the growth and the development of the fetus. During pregnancy, the mother's own nutritional status and exposure to infections have an important effect on the fetus (Fishman and others 2004). The impact of infections on malnourishment of the fetus and subsequent growth faltering has been inadequately studied (Breman, Alilio, and Mills 2004). Whereas some infectious diseases (for example, rubella) can infect the fetus through the placenta, other infections (for example, malaria and hookworm) can induce fetal death, stillbirths, and perinatal deaths, as well as contribute to poor fetal growth, without infecting the fetus (van Geertruyden and others 2004).

In addition to experiencing micronutrient deficiencies, pregnant women in developing countries are exposed to numerous environmental risks. Malaria is endemic across the tropics and subtropics, and it thrives in areas with poor drainage and stagnant water. Areas with bad sanitation provide prime conditions for hookworm infections (Hotez and others 2006). In many developing countries, and especially among the poor, malaria and hookworm infections coexist—both synergistically affecting the health of the pregnant woman and her unborn child (Watson-Jones and others 2007). Anemia in pregnancy, which is associated with increased risks of premature labor and low birth weight (Watson-Jones and others 2007), is a multifactorial condition and is often caused by a combination of malaria, hookworm infections, and dietary deficiencies (Menendez, Fleming, and Alonso 2000; Watson-Jones and others 2007; see also box 2.2).

Malaria. Every year approximately 50 million pregnant women worldwide are exposed to malaria; 30 million of these women live in the African region (Crawley and others 2007). Malarial infection during pregnancy is an important, and preventable, environmental cause of low birth weight (Allen and others 1998; van

---

BOX 2.2
Overweight Mothers Carrying Underweight Children

It is not uncommon in developing countries to see overweight mothers carrying underweight babies. These images effectively convey the message that not food security but repeated infections from poor environmental health conditions often contribute to undernutrition among children (World Bank 2006c). Moreover, food requirements during early childhood—the "window of opportunity"—are small (World Bank 2006c). Even where food scarcity is a concern, the lack of affordable environmental health services such as clean water only aggravates the situation, diverting resources from the family's food budget and thus contributing indirectly to the child's poor nutritional status (Cairncross and Kinnear 1992).

A key feature in the "fetal-programming" or Barker hypothesis is that being underweight in early infancy or even earlier, in the womb, makes the child more susceptible to rebound growth if food becomes available in abundance. Thus, those girls who were growth restricted in the womb or became stunted in early infancy could be susceptible to overweight and related chronic disease later in life (Eriksson 2005; Sachdev and others 2005). Environmental health conditions have either more distal effects (that is, the mother's own irreversibly developed underweight status during early infancy or in womb) or more proximate effects (that is, malaria and hookworms that effectively reduce birth weight). Although reduced birth weight is a strong predictor of subsequent underweight status or stunting, adverse effects of environmental health conditions on fetal growth cannot be incorporated in cohort studies that have studied the effects of infections on growth faltering (see box 2.1 and appendix A).

Geertruyden and others 2004). Malaria is thought to affect a child's birth weight in two ways: first, through placental infection, and second, through malaria-induced maternal anemia (Menendez, Fleming, and Alonso 2000). Placental infection appears to be associated with a significant reduction in birth weight and with increased risk of "low" (that is, under 2 kilograms) birth weight (Menendez, Fleming, and Alonso 2000). One review estimates that in areas where malaria is endemic, one-fifth of babies with low birth weight are underweight because of malarial infection of the placenta while the mother was pregnant (Menendez, Fleming, and Alonso 2000). Malaria also clearly contributes to maternal anemia; approximately 400,000 pregnant women develop moderate or severe anemia each year in Sub-Saharan Africa because of malaria infection (Guyatt and Snow 2004).

Hookworm infections. Hookworm infections are common worldwide but thrive in poor communities in the tropics where poor water supply and poor sanitation are common (Steketee 2003). Data from the early 1990s suggested that about a third of the pregnant women in the developing world harbored hookworm infection (Steketee 2003). Few studies have looked at the impact of hookworm infections on pregnant women (Steketee 2003). Some studies in Nepal, however, showed that hookworm infection exacerbated iron deficiency and anemia in pregnant women.

Indoor air pollution. Apart from parasitic infections such as malaria and hookworms, indoor air pollution adversely affects the health of pregnant women. Biomass is often used for cooking and heating in developing countries, leading to smoky kitchens and high levels of particulate matter. These conditions have been shown to lead to various respiratory infections in women and young children—who are the most exposed because of time spent near the stove (Bruce, Perez-Padilla, and Albalak 2002). Two studies have implicated indoor air pollution in low birth weight and perinatal mortality. A study in rural Guatemala found that birth weight was 63 grams lower for babies born in households using wood than for those born in households using cleaner fuels, and a study in Zimbabwe showed an association between perinatal mortality and exposure to indoor air pollution (WHO 2000). The association of indoor air pollution with growth faltering, however, is far less clear (Smith, Mehta, and Maeusezahl-Feuz 2004) than the association with malaria (Watson-Jones and others 2007).

Early Infancy: The "Window of Opportunity"

Several studies that looked at the impact of infections on child growth have shown that exposure to environmental health risks in early infancy leads to permanent growth faltering, lowered immunity, and increased mortality (figure 2.4; see also appendix A). Averting repeated disease episodes, especially in the first two years of life—the "window of opportunity"—prevents the more permanent and devastating wasting and stunting, which have longer-term implications for a child's health and prognosis (World Bank 2006c).

FIGURE 2.4
The Window of Opportunity for Addressing Undernutrition

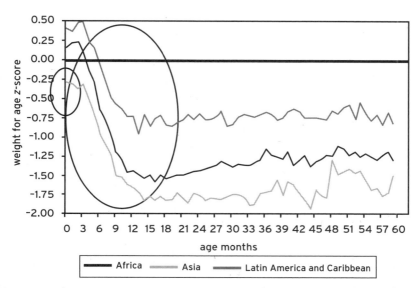

Source: Shrimpton and others 2001.

Neonates and infants up to about six to nine months of age acquire immunity from their mothers, which often prevents ARIs (Sato and others 1979). The immaturity of the infants' immune responses (Marodi 2006; Nair and others 2007; Wilson 1986), however, helps explain why early exposure to diarrheal pathogens, especially in early infancy, is so critical in the development of malnutrition. Thus, the same infections that cause disease in the developed world only among those with impaired immune systems (for example, AIDS patients) lead to diarrhea and growth faltering in infants in developing countries (Checkley and others 1998; Molbak and others 1997).

Breastfeeding is considered an effective means of protecting infants from diarrheal diseases (Dai and Walker 1999; VanDerslice, Popkin, and Briscoe 1994). The stress and fever caused by disease increase the child's food requirements while also reducing the intake and absorption of nutrients. Proper feeding practices during diarrheal episodes are critical for a child's health and recovery. In most cultures, the common practice is to restrict food from sick children, often for too long a time (Brown 2003). Only in recent years has the medical profession emphasized the need for continuous feeding during diarrheal episodes (Brown 2003). Reducing the level of environmental contamination similarly reduces the risk of diarrhea (VanDerslice, Popkin, and Briscoe 1994). Good sanitation practices protect infants by creating barriers to keep pathogens out of their environment. These practices include properly disposing of excreta to isolate human wastes, improving water supplies to protect drinking water from fecal contamination, and hand washing and personal hygiene to reduce the transmission of enteric pathogens in the home (VanDerslice, Popkin, and Briscoe 1994). Analysis of data in Bangladesh, Malaysia, and the Philippines has provided evidence that lack of breastfeeding and poor environmental sanitation have a pernicious synergistic effect on infant mortality (Habicht, DaVanzo, and Butz 1988; Pelletier 1994). The protective effect provided by good-quality drinking water and improved community sanitation is greatest for nonbreastfed infants and completely weaned infants (VanDerslice, Popkin, and Briscoe 1994). For the latter, reducing exposures (especially foodborne exposures) remains important. Strategies for reducing exposures include improvements in choice of supplemental foods, better preparation practices and storage methods, and better personal hygiene (VanDerslice, Popkin, and Briscoe 1994).

Clean Environments, Healthy Children

In early childhood (two to five years of age), the growth-faltering effects of repeated disease episodes are considered largely reversible, in contrast to the irreversibility of such effects in early infancy. Still, the environmental risk factors associated with poor access to water, improper sanitation, and bad household and community hygiene remain a threat—especially given the child's increased mobility (walking) and associated ever-increasing peri-household activities. Action should be taken at both the household level and the community level:

- *Household level.* In the late infancy and toddler periods, hand-washing practices, proper disposal of children's feces, and safe storage of milk and weaning foods are critical activities to cut diarrheal transmission. In addition, with increased mobility of the child (from crawling on the floor, to walking and running), addressing general household hygiene in terms of surfaces (including bedclothes and floors) becomes crucial, as do cleaner yards outside (soil helminths or parasites caused by lack of sanitation or animal wastes from domestic animals) (Cairncross and others 1996).

- *Community level.* Community action and control of the public domain can be seen as an important step to enable improved household and personal hygienic practices (the private domain). Both are needed, but in some cases (for example, with hookworms) collective sanitation efforts are key to success (Cairncross and others 1996). In addition, young children will benefit from indoor residual spraying or other personal-level protection against malaria even if the problem is not addressed collectively.

Averting Cognition and Learning Impacts

Cognitive function in children—reflecting an ability to learn—is affected by environmental and health-related factors (Berkman and others 2002). Risk factors that interfere with cognition are especially important in the first two years of a child's life, which marks a period of rapid growth and development (Berkman and others 2002). In early childhood, diseases attributed to environmental factors, such as diarrhea and helminth infections, also have the potential to affect a child's later cognitive functions. Over the past several years, studies have begun to investigate the impact of diarrheal illness and helminth infections during early childhood on verbal fluency, cognitive function, and school performance.

Several studies of children in Brazil and Peru have investigated the impact of diarrheal illness in the first two years of life and subsequent deficits in cognitive function as well as school performance. Among Brazilian children, the incidence of diarrheal illnesses in infancy was found to be associated with impaired cognitive performance 4 to 7 years later (Guerrant and others 1999; Niehaus and others 2002), as well as with disproportionate reduction in verbal fluency 5 to 10 years later (Patrick and others 2005). However, one large study in Peru that controlled for confounders saw no independent association with IQ at nine years of age (Walker and others 2007). The number of overt episodes of diarrhea has also been associated with indicators of school performance (Lorntz and others 2006).

Helminth infections have the highest prevalence and intensity in school-age children, and analyses of schoolchildren in Jamaica and Kenya have shown the positive effects of treatment on cognition function (Guerrant and others 1999). Only a few studies show the effect of helminths on development in children under five years of age (Walker and others 2007). In rural Nicaragua, the presence of intestinal parasites was associated with poor language performance among children (see Walker and others 2007).

In recent years, several new longitudinal studies, often enrolling close to 2,000 participants, have investigated the effect of early childhood malnutrition on educational outcomes such as school grade attainment (Alderman and others 2006; Grantham-McGregor and others 2007; Maluccio, Hoddinott, and Behrman 2006); learning productivity (Behrman and others 2006); school dropout or grade repetition (Lorntz and others 2006); and primary school enrollment (Alderman and others 2006; Lorntz and others 2006). Earlier studies had already consistently shown that an increasing level of stunting has negative effects on the studied variables (Grantham-McGregor and others 2007). Alderman and others (2006) have recently shown how one standard deviation drop in height for age leads to a 0.68-year lower school grade attainment or a 0.4-year delayed enrollment. (For a review of these studies, see appendix B.)

In addition to diarrheal illness and helminth infections, severe or cerebral malaria in children under five years of age is associated with subsequent neurological and cognitive impairments (Walker and others 2007). Data from a study of Kenyan children suggest severe *falciparum* malaria may be the single most common cause of acquired language disorder in the tropics (Carter and Mendis 2006).

Key Messages

- Malnutrition, poor environmental conditions, and infectious diseases are highly associated geographically and take their heaviest tolls on children under five years of age in Sub-Saharan Africa, South Asia, and certain countries in the Eastern Mediterranean region.
- Infections and malnutrition operate in a vicious cycle to affect child health. Until recently, however, the role of infections in the worsening of nutritional status, which, in turn, reduces growth in children, was not addressed. Over the past several decades, several cohort studies have provided strong evidence of how almost all infections influence a child's nutritional status.
- The impact of infections on child growth has shown that exposure to environmental health risks in early infancy leads to permanent growth faltering, lowered immunity, and increased mortality. In the life cycle of a child, the period from the womb to the age of about two years—the so-called window of opportunity—is critical in terms of environmental health interventions. The infections that adversely affect a child's growth have also been seen to result in cognition and learning impacts as well as chronic diseases later in life.

Note

1. The World Health Organization defines 100 liters per capita per day as the amount of water required to meet all consumption and hygiene needs. Additional information on lower service levels and potential effects on health are described in Howard and Bartram (2003).

CHAPTER 3

How Environmental Health Supplements Other Child Survival Strategies

THE BIGGEST EFFECTS ON MORTALITY can be achieved by simultaneously improving health status and nutritional status of high-mortality populations (Pelletier 1994). However, the current child survival strategies in developing countries focus mostly on case management and treatment. From the health sector's perspective, primary prevention comprises vaccinations, micronutrient supplementation, promotion of breastfeeding, and measures to decrease low birth weight, including birth spacing (Murphy, Stanton, and Galbraith 1997) (see figure 3.1). These strategies all are intended to increase the ability of the host to resist or reduce infection after exposure has occurred, but they do not attempt to reduce exposure to the environmental determinants of ill health, which constitute another aspect of primary prevention.

Additional activities relating to environmental health (sanitation, clean water, and vector control) are largely ignored because they fall outside of the current health sector mandate (see figure 3.1) even though environmental health legislation is often in place and such activities are known to be extremely effective. Within the World Bank's new Health, Nutrition, and Population strategy, the Multisectoral Constraints Assessment attempts to address this issue by looking at the multisectoral determinants of health outcomes, including those from the environmental and infrastructure sectors. The new strategy's focus on strengthening health systems—including disease surveillance systems—will also help

FIGURE 3.1
Range of Preventive Activities in Child Survival

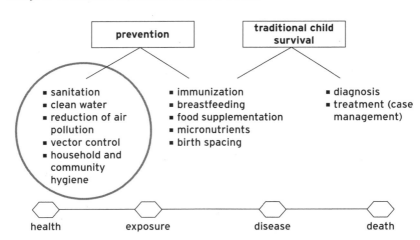

Source: Adapted from Favin, Yacoob, and Bendahmane 1999, figure 1-1.

better track environment-related infections and exposures. Given the evidence showing how environmental risks and subsequent infections affect child malnutrition, correcting the current neglect of environmental health in child survival and child health strategies in developing countries becomes even more imperative. This chapter highlights how appropriate environmental health actions can be an important complement and supplement to other existing strategies addressing child health.

Adding Value to Health Systems

Health system frameworks are designed around improving case management of childhood illnesses. Through broad frameworks such as Integrated Management of Childhood Illness, efforts are being made to strengthen case management, to benefit from synergies across vertical programs, and to capitalize on externalities in health care delivery (Mason and others 2006). Nevertheless, several areas where case management through health systems is not enough—in the case of childhood illnesses such as diarrhea, acute respiratory infections, and helminth infections, for example—call for better primary prevention of environmental risks (see table 3.1 for description of how environmental health can supplement the activities of the health sector).

With the current case management strategies of child survival programs, several issues limit the effectiveness of these interventions:

- *First* is the safety of administering drugs for parasitic infections during pregnancy. Where possible, priority should be given to treating or preventing

TABLE 3.1
Role of Environmental Health in Supplementing Health System Strategies

Cause of Death (%)	Health Systems		Role of Environmental Health
	Main Strategies	Main Limitations	
Neonatal causes (27%), omitting sepsis and diarrhea but including neonatal tetanus	Evidence-based prenatal care (tetanus toxoid important) Counseling and professional help during and after delivery	Lack of professional obstetric and prenatal care capacity	• Because of malaria and hookworms, role is potentially very large in prenatal period, which could ease pressure on scarce obstetric capacity. • Neonatal tetanus control policy may rely mainly on clean delivery (that is, on hygiene).
Acute lower respiratory infections (ALRIs) (19%)	Improved case management	Antibiotic resistance	• Role is large through indoor air pollution mitigation and becomes greater when the effects through malnutrition are considered. • There is increasing reason to believe that hygiene may play a role in preventing ALRIs.
Diarrhea (18%), including neonatal diarrhea	• Oral and intravenous rehydration therapy • Appropriate breastfeeding practices • Micronutrient supplementation	• Low compliance with oral rehydration therapy • Diarrhea incidence unaltered, but chronic and bloody diarrhea emerge as problem in an increasing proportion of the remaining deaths • Inappropriate use of antibiotics and emerging anthelminth resistance	• Role in primary prevention of diarrhea is very large through improved water and sanitation facilities and improved hygienic practices.

(continued)

TABLE 3.1
Role of Environmental Health in Supplementing Health System Strategies (*continued*)

Cause of Death (%)	Health Systems		Role of Environmental Health
	Main Strategies	Main Limitations	
		• Relatively low (about 20%) efficacy of micronutrient supplementation (especially with rampant infections) • Breastfeeding by HIV-positive mothers	
Neonatal sepsis (10%)	• Antibiotic therapy	• Antibiotic resistance and lack of access (especially in rural areas) to hospitals	• Clean delivery in household settings is augmented.
Malaria (8%)	• Case finding and case management • Bednets • Indoor residual spraying	• Antimalarial resistance • Low efficacy of bednets • Compliance issues, leading to retreatment costs	• Horizontal community involvement is needed. • Better environmental management, especially in urban areas, results.

Sources: Annual proportions for main under-five killer diseases: Bryce and others 2005; Jefferson and others 2007; Luby and others 2005; Rabie and Curtis 2006.
Note: Neonatal causes include birth asphyxia, low birth weight, and neonatal tetanus.

infections either before the first pregnancy or between pregnancies. For conditions that are diagnosed during pregnancy or that must be treated or prevented during pregnancy (for example, anemia, malaria, hookworm, or some combination of these conditions), the safety of the treatment must be balanced with the adverse consequences of the disease (Steketee 2003). Currently, reproductive health programs encourage public health approaches during pregnancy for anemia, malaria, and hookworm prevention and treatment because the substantial adverse consequences outweigh any risk associated with prevention and treatment (Steketee 2003).

- *Second*, there is growing drug resistance, as well as widespread misuse of antibiotics to treat children for infections such as diarrhea (Thapar and Sanderson 2004). In older children, the World Health Organization (WHO) is closely monitoring resistance to the drugs for helminth treatment, while problems relating to reinfection remain (Awasthi, Bundy, and Savioli 2003; Sur and others 2005). Antimalarial drug resistance is weakening the effect of chemotherapy malaria control in Africa and constraining the drug options for the early treatment strategy recommended by WHO.

- *Third* are problems related to compliance. For example, the uptake and use of oral rehydration therapy has possibly led to significant reductions in diarrhea mortality; however, use of oral rehydration therapy remains very low, at around 20 percent, in countries such as Bangladesh and India (Thapar and Sanderson 2004).

Childbirth and Infant Care Programs

The components of safe motherhood strategy usually fall primarily under the health promotion activities (care packages) organized and provided by health ministries (Graham and others 2006). They consist of, for example, community education, evidence-based prenatal care, and counseling, including primary prevention of peri- or neonatal infections (tetanus and HIV).

In many developing countries, unclean delivery practices, traditional umbilical applications, and the use of unclean cloths to swaddle the baby can have detrimental health effects on the newborn (Bennett and others 1996; Mullany and others 2006; Quddus and others 2002). Additionally, in many developing countries, the prevalence of multiresistant bacteria that cause neonatal infections (a major cause of neonatal deaths) is rapidly increasing, even while broad-spectrum antibiotics are losing their effectiveness (Bang and others 1999; Vergnano and others 2005). Also, difficulties remain in maintaining high coverage of tetanus immunizations for pregnant women (to prevent neonatal sepsis). In these areas, hygienic and clean delivery practices, supplemented by environmental health interventions approaches, become even more important (Mullany and others 2006; Quddus and others 2002).

By advocating hygienic practices for delivery and infant care, several health sector programs in developing countries are already adopting environmental health principles related to clean household hygiene. Additional environmental health interventions, such as providing clean water and sanitation, play an augmentive—yet important—role in these health promotion efforts. Only with adequate quantities of water and proper sanitation facilities can households with infants maintain clean surfaces, practice proper hand washing, and properly dispose of feces.

Another important tenet of infant care programs is appropriate feeding practices. Established evidence shows that exclusive breastfeeding can prevent malnutrition (Adair and others 1993; Alvarado and others 2005; Chisti and others 2007; Cohen and others 1995). Consequently, infant care programs through the health sector are encouraging breastfeeding practices as well as the postponement of the weaning period among severely malnourished infants to prevent death (Pelletier 1994). Additionally, as mentioned earlier, inadequate sanitation compounds poor breastfeeding practices and synergistically affects child health. This need for improved sanitation to reduce diarrheal transmission to infants is seen even more starkly among countries with high rates of HIV prevalence. Medical research has shown that for HIV-positive mothers, as many as 9.5 percent of their infants can acquire HIV through breast milk over a three-year period (Bertozzi and others 2006). These mothers are faced with a very difficult dilemma—whether to breastfeed or not—because their children may face severe malnourishment (caused by diarrhea from bad water and poor sanitation) or alternatively might acquire HIV if breastfed. Exclusive breastfeeding, even though uncommon, might be safer than mixed breastfeeding for reducing HIV transmission risk (Coovadia and others 2007). Even in these circumstances, however, healthy environments remain a high priority.

Immunization Programs

The global Expanded Program on Immunization has been very effective in reducing mortality in children under five years of age (Brenzel and others 2006)—with sustainable reductions especially through measles, whooping cough (pertussis), and tetanus vaccines (Brenzel and others 2006). There is still potential to save an additional 1.8 million children under the age of five annually by improving coverage with current licensed vaccines in South Asia and Sub-Saharan Africa (Brenzel and others 2006); however, economic costs as well as logistical considerations (for example, the need to maintain a cold chain for live vaccines) constrain this possibility. A very recent study from the WHO and United Nations Children's Fund, reporting a 91 percent drop in the number of measles deaths in Africa, has attributed the decrease largely to better coverage in routine immunization programs and targeted campaigns (Zaracostas 2007).

In recent years, a number of new vaccines against rotavirus, pneumococcal infections, hookworms, and malaria have been developed and tested. The rotavirus vaccine has shown promise but still faces several technical hurdles before it can be used widely against diarrheal disease in developing countries (Banerjee and others 2006; Thapar and Sanderson 2004). Also important, rotavirus—unlike other intestinal pathogens—apparently is not strongly related to the lack of clean water, sanitation, and appropriate hygiene (Thapar and Sanderson 2004), perhaps because a low number of viruses is needed to infect an individual and because the virus tends to survive outside a host a long time—that is, it is difficult to "wash away" (Eisenberg and others 2006).

Some progress has also been achieved in hookworm vaccines (Chu and others 2004), and several malaria vaccines are currently being tested in clinical trials, where they have shown to provide some protection (Komisar 2007). Although progress continues, with some setbacks (Cutts and others 2005; Singleton and others 2007), in the field of new vaccines (for example, bacterial vaccines against neonatal sepsis and meningitis or pneumococcal infections and diarrhea for children under five), the improvement of environmental health conditions—especially water and sanitation—should continue to be emphasized (Thapar and Sanderson 2004).

Micronutrient Supplementation Programs

Along with underweight, micronutrient deficiencies—such as in vitamin A, iron, and zinc—contribute to the global burden of disease in young children. Like underweight, micronutrient deficiencies increase susceptibility to infections, which, in turn, may further aggravate the deficient state (West 2003).

A number of micronutrient supplementation programs are being undertaken in developing countries. Several studies have shown that vitamin A and zinc play key roles in determining the duration, frequency, and severity of infectious diseases such as diarrhea (Guerrant, Lima, and Davidson 2000). However, evidence also shows that the burden of infections, in turn, often constrains the effectiveness of these supplementation programs. Likely, this effect is caused by malabsorption of key antimicrobial drugs, resulting in the emergence of drug resistance in impoverished areas (Guerrant, Lima, and Davidson 2000).

Research has revealed that infectious diseases, such as diarrhea and respiratory infections, and intestinal worm burdens may modify the growth response to vitamin A supplementation (Hadi and others 1999; West 2003). With a heavy burden of respiratory infections, supplemental vitamin A may not be completely absorbed and can be expected to be excreted in high quantity (Hadi, Dibley, and West 2004). In a research trial of 1,405 Indonesian children, those who experienced respiratory infections for more than 26 days during a four-month observation period had a 74 percent reduction (0.17 centimeters in four months) in their linear growth response to vitamin A supplementation compared with children with no respiratory infection (Hadi and others 1999). Results from trials of iron supplementation have not found significant growth effects, even in anemic children (Iannotti and others 2006). Iron supplementation, in addition, may actually lead to increased morbidity and consequently to reduced dietary intakes and poor nutrient absorption (Iannotti and others 2006). Also, studies have found that the degree of infectious disease treatment is an important factor in determining who benefits and who is harmed in terms of serious infectious disease outcomes when low-dose iron supplements are provided (Iannotti and others 2006).

Interventions to prevent or decrease malnutrition and infectious disease are expected to decrease child mortality, and interventions that accomplish both will have the greatest effect (Pelletier 1994; Pelletier and Frongillo 2003). Environmental health interventions have an important complementary as well as supplementary role to play in addressing nutritional deficiencies. Water, sanitation, and hygiene have the potential to reduce zinc deficiencies that occur from losses during diarrheal illnesses. Similarly, efforts to reduce or control respiratory infections will help increase the effectiveness of vitamin A supplementation programs in improving growth in children (Hadi, Dibley, and West 2004).

Adapting Environmental Management Programs

Recent evidence from verbal autopsy data provides concrete evidence that the mortality rates of children under five years of age caused by malaria in Sub-Saharan Africa remain very high (Adjuik and others 2006). Key strategies in the control of malaria have relied on early diagnosis and treatment, on the use of insecticide-treated nets (ITNs), and—to a lesser extent—on indoor residual spraying (Mabaso, Sharp, and Lengeler 2004; WHO 2003). The gains in malaria control in Sub-Saharan Africa achieved in the 1940s to 1970s from indoor residual spraying with DDT (dichloro-diphenyl-trichloroethane) and other insecticides were eroded in the mid-1980s—with malarial epidemics becoming frequent and more severe (Mabaso, Sharp, and Lengeler 2004).

ITNs have been shown to reduce all-cause mortality by about 20 percent in community-wide trials in several African settings (WHO 2003), although a recent observational community-based intervention study suggests a substantially higher efficacy of about 44 percent (Fegan and others 2007). Poor compliance with net retreatment, combined with the additional cost of the insecticide and a lack of understanding of its importance, however, pose difficult barriers to full implementation of ITNs in endemic countries (Biran and others 2007; National Center for Infectious Diseases 2006; WHO 2003). In the case of communities with longer-lasting nets, the problem of retreatment is more or less obsolete, but the problem of adherence to best practices with use of these nets remains an issue.

With few options for malaria, many rural households in Sub-Saharan Africa continue to rely on smoke from indoor cooking as a mosquito repellent (Kristensen

2005). A recent review of research evidence investigated the association between indoor air pollution and malaria transmission (Biran and others 2007). The main findings were that smoke from domestic fuel use probably does not have much effect on mosquito feeding, while the presence of eaves spaces makes houses more permeable to mosquitoes and has been associated with an increased risk of malaria (Biran and others 2007). Soot from domestic fires may impair the effectiveness of ITNs because the nets are perceived to be dirty, possibly resulting in an increase in the frequency of washing them and a consequent loss of insecticide (Biran and others 2007). Conversely, it can be concluded that a reduction in indoor air pollution might have beneficial effects on malaria control by allowing ITNs to remain unwashed for longer periods.

Strong evidence indicates that environmental management has been effective in controlling malaria, although mosquito ecology in many tropical areas often makes this approach impractical (Fewtrell and others 2007; Keiser, Singer, and Utzinger 2005).[1] Malaria control methods that involve environmental management—that is, a modification or manipulation of the environment to reduce malaria transmission (for example, through the installation, cleaning, and maintenance of drains or the systematic elimination of standing pools of water)—currently receive far less attention (Keiser, Singer, and Utzinger 2005).

Adjusting Infrastructure Strategies

Infrastructure investments (such as in water supply, sanitation, and rural energy) are typically designed with the primary objective of improving access or efficiency rather than maximizing public health effects, although such interventions do sometimes recognize the importance of public health. Yet with growing emphasis on increasing access to water supply systems and sanitation facilities and on moving poorer populations up the energy ladder, there remains a huge opportunity to appropriately incorporate public health considerations into the design and delivery of these interventions.

Improving Water Supply

More than 1.1 billion people still do not have access to safe drinking water. Service is poor even in many countries that have water supply systems. For many consumers, piped water is often intermittent, and when available, it is unsafe for drinking. With growing demands for water, governments in developing countries are making some progress in extending coverage to improved water supply[2] with large-scale water supply projects, especially in urban areas.

From a health perspective, water quantity is generally more important than water quality, because increased quantities of water promote good hygiene and can prevent fecal-oral transmission by a number of different routes; increased quantities of water also reduce skin and eye infections (Boeston, Kolsky, and Hunt

2007). Furthermore, household water connections are considered optimal because consumption of water doubles or triples as households use it for improved hygiene practices (Cairncross and Valdmanis 2006), and far greater health benefits result.

In both urban and rural areas in developing countries, people demand water supplies to reduce the time and cost of fetching water. To meet demands for water, governments have been increasing efforts to extend the coverage of piped water systems. The major investments in urban water focus on network solutions and thus reflect the needs of utilities, which are concentrated on providing satisfactory service to their existing customers; managing large amounts of water safely and efficiently; meeting environmental standards for their wastewater discharges; and recovering costs to permit continued operation, maintenance, replacement, and expansion. Accordingly, engineering, financial, and environmental perspectives dominate the design of network interventions, and public health is reflected only in technical standards to reduce the risk of common-source outbreaks from the water supply (Boeston, Kolsky, and Hunt 2007).

In rural areas, the time-saving benefits of water access are even more pronounced, and the "most obvious benefit is that water is made available closer to where rural households need it" (Cairncross and Valdmanis 2006: 773). Accordingly, government investments in rural water supply aim at improving access; however, the smaller size of rural communities means that piped systems, in general, and house connections, in particular, tend to be more expensive per capita (Cairncross and Valdmanis 2006).

Microbial contamination of drinking water in rural and urban areas can occur either at the source, through seepage of contaminated runoff water, or within the piped distribution system. Moreover, unhygienic handling of water during transport or within the home can contaminate previously safe water (WHO 2007c). Household water treatment technologies (such as chlorination and solar disinfection) and safe storage can be important interventions in situations where access to water supplies is secure but household water quality is not assured (WHO 2007c).

Improving Sanitation

Roughly 2.6 billion people in the developing world are without adequate sanitation,[3] and facilities are often overloaded, in disrepair, or unused. Even though the sanitation gap is twice as large as that of water supply, investments in sanitation and hygiene have lagged far behind those in water and other "social" sectors, such as health and education.

The main costs of urban sanitation services are those of sewers and sewage treatment. Whereas sewers contribute to public health through reducing everyday contact with sewage (especially by children), wastewater treatment is designed largely to meet ecological objectives and not those of public health (Boeston,

Kolsky, and Hunt 2007). Urban utilities, by and large, are not well designed or staffed to address off-network solutions for water supply or sanitation, yet those solutions are likely to be the most important first steps of progress in environmental health for many of the urban poor. Rural sanitation is frequently a low priority relative to food, shelter, schooling, and employment. Families that invest in rural sanitation usually do so for other reasons—for cleanliness, dignity, modernity, comfort, and convenience—more than for public health.

For a variety of historical, political, and institutional reasons, public health plays little role in project design, despite a large role in project justification of both water supply and sanitation investments in developing countries (Kolsky and others 2005). The roles for the infrastructure sector have largely been related to improving access (for example, increasing the number of tap connections), while those of the health sector have traditionally focused on improved treatment strategies. Opportunities remain for the health sector to take on a larger role in improving the health outcomes of infrastructure investments through promotion (for example, hygiene promotion or support to the marketing of sanitation) and advocacy (for example, persuading utilities to lower their off-putting connection charges and to recover the connection costs through the monthly tariff so that house connections are more affordable to the poor). Relatively easy improvements in public health outcomes can be achieved in a number of ways:

- Relatively low-cost hygiene promotion programs that attempt behavior change in household hygiene through a wide range of social-marketing tools can be implemented. Neither network utilities nor rural water engineering staff members are naturally inclined to assign high priority to hygiene promotion, although it is highly cost-effective as a health intervention (Laxminarayan, Chow, and Shahid-Salles 2006).
- Urban water and sewerage network expansions can often be sequenced in different ways, and a relatively easy way to speed public health improvement is to stress the need to extend services to new areas as quickly as possible.
- Trade-offs between ideal environmental objectives and the priorities of public health can be explored. Wastewater treatment systems are, by and large, designed to achieve ecological objectives, not public health ones. If strict environmental standards are adopted in poor countries, the often high costs of such systems means that wastewater treatment comes at the expense of depriving the poor of basic access to sanitation. The costs and benefits of wastewater treatment need to be balanced against the public health benefits and costs of basic access.

Moving up the Energy Ladder

Worldwide, more than 3 billion people continue to rely on solid fuels, including biomass and coal, to meet their most basic energy needs: cooking, boiling water, and heating (WHO 2006). To address this issue, developing countries have made various rural energy investments to move these people up the energy ladder toward cleaner fuels. Nevertheless, progress since 1990 has been negligible because the small gains made are lagging population growth (WHO 2006).

Recent decades have witnessed many household energy initiatives, ranging from ambitious government-run improved cookstove programs to small-scale community-led projects. Cleaner fuels and improved cookstove programs that reduce indoor air pollution can mitigate diseases such as respiratory and other illnesses (World Bank 2007d). Studies in India and Nepal have revealed that women who were exposed to biomass smoke—but who did not smoke themselves—had chronic respiratory disease death rates comparable to those of heavy male smokers (Modi and others 2005; World Bank 2007d).

Improved cookstove programs have an enormous potential to affect respiratory health. They can especially benefit women and young children, who are disproportionately affected by respiratory disease. In the past, however, such programs have more often focused on energy-efficiency criteria rather than on health considerations. The first improved cookstove programs in India, for example, were more concerned with the potential to burn biomass energy more cleanly and efficiently than with health outcomes. Secondary benefits generally included conserving fuel, reducing smoke emission in the cooking area, keeping kitchens clean, reducing deforestation (Agarwal 1983, 1986; World Bank 2007d), lessening the drudgery of women, reducing cooking time, and improving employment opportunities for the rural poor. The most organized intervention of rural household stoves—China's National Improved Stove Program (NISP)—focused primarily on fuel efficiency. However, a major independent review of NISP, including a 3,500-household survey in three provinces at different levels of coverage, found that the program improved air quality and health in rural households, but not sufficiently to meet Chinese indoor air quality standards (Zhang and Smith 2005).

Whereas the supply of technology is more concerned with energy efficiency and costs, the demand for cookstoves has reflected limited awareness of the health effect of improved cookstoves. In India, for example, perceived benefits cited by users of improved cookstoves included reduction in time spent in cooking, reduced fuel consumption, health benefits such as reduction of eye irritation, and cleaner kitchens. Yet only 6 percent of the households cited "health benefits" as a perceived benefit. This finding suggests that users over the years had accepted the burning sensation in the eyes and coughing as daily facts of life and considered only medical issues that required a visit to the rural doctor as "health" issues (World Bank 2007d). In Guatemala, similarly, respondents from three improved cookstove programs were found to value the benefits from lower fuelwood use, reduced cooking times, and less smoke. Significantly, the advantages of reduced eye irritation and better respiratory health were perceived to be the lowest benefits under the program (Ahmed and others 2005). This evidence of households undervaluing the health benefits associated with improved cookstoves highlights opportunities for raising awareness through a variety of public media or social-marketing approaches.

From a health perspective, behavior-change components are also important supplements in clean stove programs. In Bangladesh, for example, pollution exposure for children in a typical household can be halved by adopting two simple measures: increasing children's outdoor time from three to five or six hours a day and concentrating outdoor time during peak cooking periods (Dasgupta and others 2004b). Including such health promotion activities in rural energy projects is important in raising awareness of health benefits associated with improved indoor air and creating a demand for improvement of cookstoves and changes in behavior.

Key Messages

- Current child survival strategies in developing countries focus on case management and treatment, to the neglect of primary prevention—especially relating to reducing exposure to infections. These strategies typically intend to increase the ability of the host to resist or reduce infection after exposure has occurred, but they do not attempt to reduce the exposure to environmental determinants of ill health.

- Environmental health actions incorporated into existing child health strategies can provide a bigger "bang for the buck" in terms of health improvements, and they are an important complement and supplement to other existing strategies addressing child health by adding value to health systems, adapting environmental management programs, and adjusting infrastructure strategies.

Notes

1 Conversely, a neglect of environmental health action through vector habitat control has been responsible for the resurgence of dengue virus in almost all continents (Cattand and others 2006; Kay and Nam 2005) and a forceful comeback of African trypanosomiasis spread by tsetse flies in Central and East Africa in the past few decades (Remme and others 2006).

2 *Improved water supply* is defined to encompass reasonable access to protected water resources, which include rainwater collection, protected springs and dug wells, boreholes, public standpipes, and household connections. In addition, it involves the application of measures to protect the water source from contamination (Hutton and Haller 2004). *Reasonable access* means at least 20 liters per person per day, accessible within 1 kilometer of that person's dwelling (WHO and UNICEF 2005).

3 Improved sanitation involves access to sanitation facilities that allow for safe disposal of excreta, which may include connection to a sewer or septic tank system, pour-flush latrine, simple pit, or ventilated improved pit latrine (WHO and UNICEF 2005).

PART II

Economics

CHAPTER 4

How Large Is the Environmental Health Burden?

THE PREVIOUS CHAPTERS REVISITED THE RESEARCH evidence on malnutrition-infection links to demonstrate the importance of environmental health interventions for child survival and development and pointed toward the complementary role of environmental health interventions in other child survival strategies. Given this evidence, quantitatively measuring the importance of environmental health issues for better child health becomes important. Furthermore, to make decisions regarding investments in environmental health actions, policy-makers need to get a sense of the averted health costs. Hence, this chapter seeks to answer the following questions:

- How large is the environmental health burden for children under five, after considering its links through malnutrition?
- Which subregions have the highest environmental health burdens?
- What are the economic costs associated with these burdens at a country level?
  Answering these questions will help policy-makers appreciate the magnitude of the environmental health burden and better integrate environmental health into decision making for economic development. Furthermore, translating health effects into economic costs—such as a percentage of a country's gross domestic product—has been a powerful means of raising awareness about environmental health issues and facilitating progress toward sustainable development.

This chapter looks at the methodology used by the World Health Organization (WHO) to estimate the "burden of disease" within and across various subregions of the world, specifically in the context of environmental health risks. This methodology—dynamic and evolving over the past decade—has been applied to assess and compare the health status of populations in countries and subregions. Results from the latest WHO analysis of environmental health burdens incorporate the complex links between malnutrition and infections into the calculations and provide a basis for comparisons across different subregions. The next chapter extends this analysis to estimate the corresponding burden at the country level and the consequent economic costs to society.

Burden of Disease

In the field of health, a challenge one faces when trying to choose between different health sector interventions is deciding how to compare different health outcomes (or end points). For example, how does one compare environmental interventions that differ in terms of how many lives of children are saved (mortality) and how many days of various illnesses (morbidity) are avoided? How would one choose between an intervention that saved 20 child lives and 200 cases of illness, for example, and another that saved 15 lives and 1,000 cases of illness? Also, how would one compare a case of diarrheal illness averted to a case of respiratory infection averted? The Global Burden of Disease Project, initiated by WHO in 1996, aims at answering these questions and, in doing so, provides useful guidance for policy making (WHO 2007a).

To assist governments in setting priority actions in health, WHO developed a methodology to evaluate the burden of disease and to quantify the health status of a population. One of the most commonly used summary measures of population health is the disability-adjusted life year (DALY), which creates a common metric to measure both the potential years of life lost to premature death and the years of healthy life lost because of disability or illness. One DALY can be thought of as one lost year of healthy life.

Environmental Burden of Disease

A 1999 paper assessing the global environmental burden of disease reported that about 43 percent of the total burden of disease attributable to environmental risks falls on children under five years of age, even though they make up only 12 percent of the population (Smith, Corvalán, and Kjellström 1999).[1] Since then, important reports on environmental health burden estimation were undertaken in the framework of the Global Burden of Disease Project. In 2002, a section of the *World Health Report* (WHO 2002: chapter 4) quantified the significance of environmental health risks, which were revisited in depth in 2004 by the Comparative

Risk Assessment project[2] (Ezzati and others 2003; Ezzati, Lopez, and others 2004; Ezzati, Rodgers, and others 2004; Ezzati, Vander Hoorn, and others 2004).

The method used to calculate the environmental burden of disease is based on an exposure approach, supported by a comprehensive analysis of evidence on specific health risks. Exposure-response relationships for a given risk factor—such as poor water, sanitation, and hygiene (WSH)—are obtained from epidemiological studies, and the derived attributable fractions are then applied to disease burden (in deaths or DALYs) associated with the same risk factor (see WHO n.d.). WHO's (2004) Comparative Risk Assessment considers the following six environmental risk factors and examines them in relation to the diseases identified in table 4.1 (Ezzati, Lopez, and others 2004). The risk-factor approach quantified WSH and indoor air pollution as having the greatest effect on children under five years of age (Ezzati, Vander Hoorn, and others 2004; Ezzati, Rodgers, and others 2004). This risk-factor approach to assessing health status is particularly useful to Policy-makers because it enables them to think of specific interventions to address specific environmental risks.

In 2006, WHO published Prüss-Üstün and Corvalán's environmental health report, *Preventing Disease through Healthy Environments*. Using a different approach from earlier studies, the report elicited expert opinions on the fractions of different diseases attributable to environmental factors. Apart from water, sanitation, and indoor air pollution, malaria emerged as an important environmental health problem among children under five—especially in Sub-Saharan Africa. The report estimated that among children up to 14 years of age, the proportion of deaths attributable to the environment was as high as 36 percent (Prüss-Üstün and Corvalán 2006), which compares to an attributable disease burden of 24 percent

TABLE 4.1
Environmental Risk Factors and Related Diseases Included in the Comparative Risk Assessment

Risk Factor	Related Diseases
Water, sanitation, hygiene	Diarrheal diseases, trachoma, schistosomiasis, ascariasis, trichuriasis, hookworm disease
Indoor air pollution	Chronic obstructive pulmonary disease, lower respiratory infections, lung cancer
Outdoor air pollution	Respiratory infections, selected cardiopulmonary diseases, lung cancer
Lead	Mild mental retardation, cardiovascular diseases
Climate change	Diarrheal diseases, malaria, unintentional injuries, protein-energy malnutrition
Selected occupational risk factors	Unintentional injuries, hearing loss, cancers, asthma, chronic obstructive pulmonary disease, lower back pain

Source: WHO 2002.

for all age groups. A number of earlier studies had led the way toward these important findings.

In the same report, expert opinion indicated that poor WSH contributed to 50 percent of the consequences of childhood and maternal underweight (see box 4.1). For example, when environmentally attributable deaths from acute lower respiratory infection (ALRI) were reported, they were caused by indoor air pollution. Nevertheless, perhaps up to 50 percent of these deaths would not have occurred if the child had not been malnourished (Ezzati, Vander Hoorn, and others 2004). The implications of this link were mainly reported through the consequences of different environmental risks as deaths in different disease categories. Current literature on risk factors has internalized a significant portion of these consequences of malnutrition, often without much discussion. Although the WHO report (Prüss-Üstün and Corvalán 2006) did include a relatively small category (measles and meningitis deaths) of child deaths associated with

---

BOX 4.1
Why 50 Percent? Supporting Evidence from Recent Cohort Studies

In *Preventing Disease through Healthy Environments* (Prüss-Üstün and Corvalán 2006), expert opinion was cited as indicating that poor water, sanitation, and hygiene and inadequate water resources management contributed to 50 percent of the consequences of childhood and maternal underweight. This expert opinion builds on evidence that emerged toward the end of the 1950s on the impacts of environmental infections on a child's growth (Keusch 2003; Scrimshaw, Taylor, and Gordon 1959).

In this report, a technical review of 38 recent cohort studies further corroborates this 50 percent (confidence interval 39 to 61 percent) figure used to estimate the consequences of malnutrition attributable to environmental risk factors. Evidence from several of the studies demonstrates how exposure to environmental health risks in early infancy leads to permanent growth faltering, lowered immunity, and increased mortality. Using Bradford Hill's causality criteria as a framework, the results from the studies were scrutinized to evaluate whether infections cause growth faltering and, if so, to what extent (for details, see appendix A).

For example, a recent relatively large study from Bangladesh reveals that dysentery and watery diarrhea together could retard weight gain by 20 to 25 percent compared with those periods where no infections occurred (Alam and others 2000). This weight gain retardation is likely to be significantly higher when compared with international standards, and simulations that reveal 35 percent weight gain retardation result in approximately similar environmentally attributable health burden as the estimate in Prüss-Üstün and Corvalán (2006).

malnutrition, it did not include other deaths that would have been avoided had the child not been malnourished.

Including the Malnutrition-Infection Link

As discussed in chapter 2, there is now substantial evidence that repeated infections in early childhood—mainly diarrheal disease from inadequate WSH—is a significant contributor to child malnutrition (Prüss-Üstün and Corvalán 2006). Moreover, the evidence indicates that malnutrition, in turn, increases the risk of child mortality and the incidence of morbidity[3] and impairs cognitive development with consequences for educational attainment and performance (see Fishman and others 2004).

A WHO analysis in 2007 sought to incorporate this evidence of malnutrition-infection links in its environmental burden of disease methodology (Fewtrell and others 2007). For the first time, this methodology not only incorporated direct health risks from environmental factors (such as diarrheal disease burden from poor water and sanitation) but also sought to include indirect risks (concentrating on WSH and its indirect effect on mortality through malnutrition). Thus, while a traditional burden of disease calculation would associate WSH with only diarrheal diseases (see figure 4.1), the inclusion of the indirect path implies the need to include all diseases attributable to malnutrition (because 50 percent of the consequences of malnutrition are, in turn, attributed to poor WSH).[4]

These additional diseases include ALRIs, measles, malaria, protein-energy malnutrition, and other infectious diseases that cannot be directly attributed to environmental risk factors. Figure 4.1 illustrates these direct and indirect effects. To quantify the health effects of malnutrition, Fewtrell and others (2007) adopted the methodology of Blössner and de Onis (2005), who estimated fractions attributable to malnutrition for diarrheal diseases, low birth weight, lower respiratory infections, malaria, measles, and protein-energy malnutrition.

Note that to avoid possible double-counting, Fewtrell and others (2007) did not include as an indirect effect diseases such as diarrhea that are a direct effect of poor WSH, even though diarrhea is one of the consequences of malnutrition. Therefore, the estimates in Fewtrell and others (2007) are conservative.

Environmental Health Burdens

Results from the latest WHO analysis (Fewtrell and others 2007) show the water-related (WSH and water resource management, or WRM) environmental health burden for children under five years of age in various subregions (see figure 4.2). From a point of view of child health, three subregions in Sub-Saharan Africa and South Asia stand out in terms of the environmental health burden of disease.

An interesting feature of figure 4.2 is that regions differ widely from each other depending on whether one looks at DALYs per capita (panel a) or total DALYs

FIGURE 4.1
The Health Effects of Environmental Risks Factors

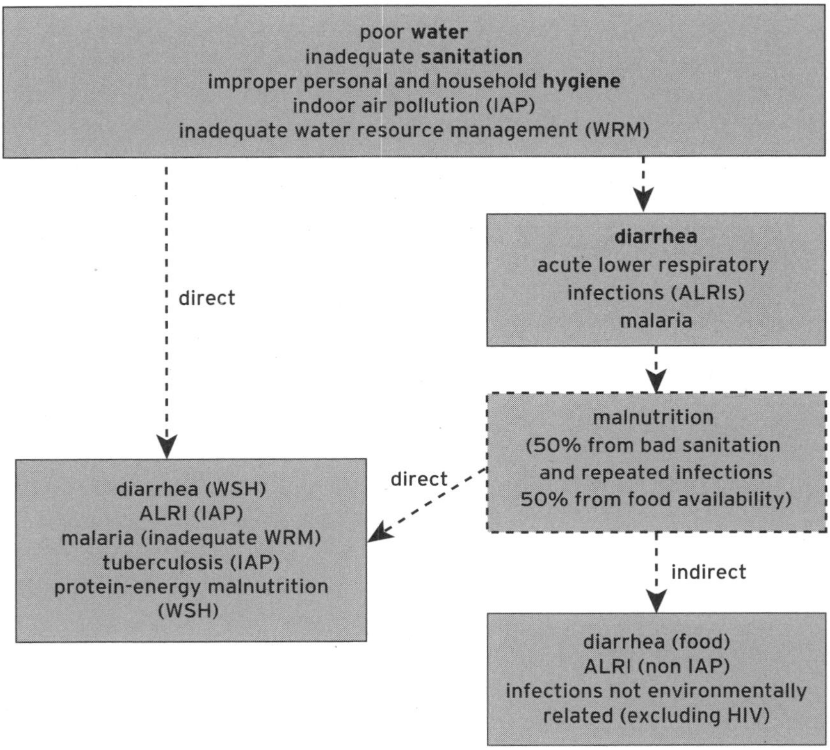

Source: Compiled by World Bank team.

(panel b). Although the Southeast Asian subregion SEARD has a similar number of DALYs as the African subregion AFRD, the two subregions differ quite substantially in terms of DALYs per capita—with the latter having a value more than two times higher. Incidentally, diarrheal disease in SEARD is, in relative terms, more important as a percentage of total DALYs than diarrheal disease in AFRD. When compared with the most developed subregion, EURA, which is in Europe, the environmental health burden (in DALYs per 1,000 children under five) in the two subregions of Africa (AFRB and AFRD) is about 150 to 175 times higher. For SEARD, this ratio is about 70 times higher. Important risk factors such as indoor air pollution and environmental-attributable perinatal mortality are not included in figure 4.2.

In the public's imagination, the home of the malnourished child is Sub-Saharan Africa. Nevertheless, some researchers showed evidence that the

FIGURE 4.2
Water-Related (WSH plus WRM) Burden of Disease in Children under Five Attributable to Environmental Risk Factors by WHO Region, 2002

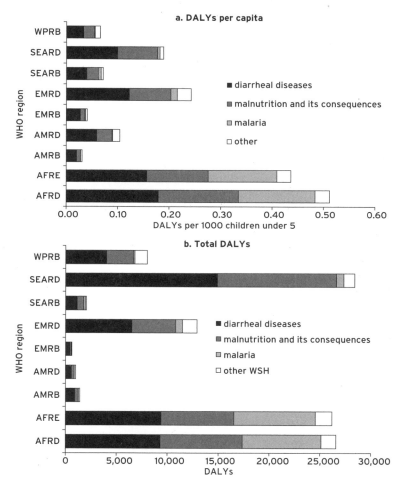

Sources: Adapted from Fewtrell and others 2007; Prüss-Üstün and Corvalán 2006.
Note: Calculations are based on deaths attributable to perinatal conditions published in Graham and others (2006: 507) and on estimating the regional aggregate number of live births using live birth per 1,000 data population data from the World Bank Development Data Platform and Live Database time series for 2000. The WHO regions are defined as follows: WPRB (Western Pacific subregion with low child and low adult mortality); SEARD (Southeast Asian subregion with high child and high adult mortality); SEARB (Southeast Asian subregion with low child and low adult mortality); EMRD (Eastern Mediterranean subregion with high child and high adult mortality); EMRB (Eastern Mediterranean subregion with low child and low adult mortality); AMRD (subregion of the Americas with high child and high adult mortality); AMRB (subregion of the Americas with low child and low adult mortality); AFRE (African subregion with high child and very high adult mortality); AFRD (African subregion with high child and high adult mortality). For a detailed list of the countries that belong to each subregion, see http://www.who.int/choice/demography/regions/en/index.html.

worst-affected region is not Africa but South Asia (Ramalingaswami, Jonsson, and Rohde 1996). Over the past few decades, other explanations have been suggested for these surprising results, which have been labeled the "Asian enigma" (see box 4.2).

BOX 4.2
Revisiting the "Asian Enigma"

As many as 30 percent of children in India are born with low birth weight, as compared with only 16 percent in Sub-Saharan Africa. Furthermore, continuing up to five years of age, underweight rates remain far lower in Sub-Saharan Africa (approximately 30 percent) compared with those in South Asia (close to 50 percent). Although the depth of hunger is clearly greatest in Sub-Saharan Africa (FAO 2006), and the gross domestic product and living standards are higher in South Asian countries, this nutritional status difference is considered to be somewhat of an enigma (Gragnolati and others 2006).

Using multivariate analyses of household survey data, Ramalingaswami, Jonsson, and Rohde (1996) suggested that this difference was explained by the better decision-making power of women in Sub-Saharan Africa compared with those in Asia, but they also stated that sanitation differences (use of latrine) between the regions clearly contribute to the nutritional status gap as well.

Scientific inference from household surveys is often hampered by issues of confounding (other related factors ultimately may be the culprit) as well as by inconsistencies in definitions (what is meant by *latrine* in different countries?) (Cairncross and Kolsky 1997). Also, for example, high latrine use might significantly threaten the integrity of sources of drinking water—again making inferences difficult (Kimani-Murage and Ngindu 2007). Often, serious problems with data quality exist; for example, in a recent Kenya National Census, the enumerators missed Kibera, the largest shantytown in Africa, which led to overestimating latrine use in Kenya. Finally, mechanisms of causality often remain obscure in studies relying on indicator data (Cairncross and Kolsky 1997; Kolsky and Blumenthal 1995). If sanitation and hygiene conditions were poorer in South Asia, one might logically expect more diarrhea per child—which is not the case (see figure 4.2).

In attempting to explain this enigma, researchers have not adequately considered the intergenerational dimensions of malnutrition—that is, malnourished fetuses result in babies with low birth weight. These babies become underweight mothers and, in turn, give birth to more babies with low birth weight. The prevalence of fetal undernutrition (or restricted growth in utero) is directly related to the duration and incidence of malnutrition (which is increased by mother's low weight before pregnancy and by infections during pregnancy, such as hookworms and malaria).

(continued)

A number of miscarriages, stillbirths, and perinatal deaths (owed in large part to rampant malaria) likely are preferentially harvesting the underweight fetuses and children from the African population (Adjuik and others 2006; van Geertruyden and others 2004). These early deaths, then, do not add to the statistics of underweight children, and therefore the region appears to have fewer underweight babies than does South Asia.

The Multiplier Effect for Environmental Health Interventions

Results from these subregional data show the multiplier effect of environmental health interventions on mortality—where investments addressing environmental risks (for example, lack of water and sanitation) not only reduce diarrhea mortality but also reduce mortality from malnutrition-related diseases and its consequences on educational attainment. In these subregions, this multiplier ranges from 1.5 to 1.9 (figure 4.3), which means for every direct death avoided

FIGURE 4.3
Mills-Reincke Ratios for Subregions

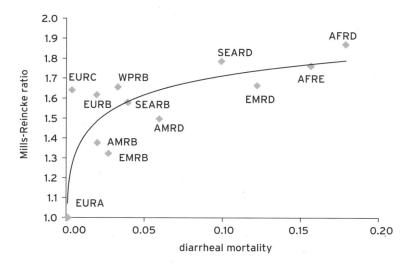

Source: Compiled by World Bank team on the basis of Fewtrell and others 2007.
Note: The World Health Organization regions are defined as follows: WPRB (Western Pacific subregion with low child and low adult mortality); SEARD (Southeast Asian subregion with high child and high adult mortality); SEARB (Southeast Asian subregion with low child and low adult mortality); EMRD (Eastern Mediterranean subregion with high child and high adult mortality); EMRB (Eastern Mediterranean subregion with low child and low adult mortality); AMRD (subregion of the Americas with high child and high adult mortality); AMRB (subregion of the Americas with low child and low adult mortality); AFRE (African subregion with high child and very high adult mortality); AFRD (African subregion with high child and high adult mortality); EURA (Europe subregion with very low child and very low adult mortality); EURB (Europe subregion with low child and low adult mortality); EURC (Europe with low child and high adult mortality).

from an environmental health intervention, at least half to almost one additional death is averted from indirect links. A recent research paper using household survey data for rural China similarly found evidence of a 2.0 multiplier, implying that for every diarrheal death averted in children under five from household water connections, an additional ALRI death was also averted (Jacoby and Wang 2004). This multiplier effect is not new: it was first empirically observed in the late 19th century and came to be known as the Mills-Reincke phenomenon (see box 4.3).

BOX 4.3
The Mills-Reincke Phenomenon

In the late 19th century, impressed by the beneficial effects of clean water supply on mortality (especially child mortality), a German sanitary engineer (Johann Julius Reincke) and an American sanitary engineer (Hiram F. Mills) independently found that for every prevented death from waterborne diseases, additional pulmonary tuberculosis, pneumonia, and infant deaths were averted. This curiosity came to be called the *Mills-Reincke phenomenon* (Sedgwick and Macnutt 1908).

A recent analysis of several U.S. cities for 1900 to 1936 has conservatively estimated that, for one diarrhea death, three other deaths were averted by clean water supply systems (Cutler and Miller 2005). Mortality declined steeply without major advances in curative and individually targeted preventive interventions (for example, vaccines) (Cutler and Miller 2005). Another important source of typhoid fever, killing especially children under five years of age from unclean milk, was also eliminated in the early decades of the 21st century (Rosen 1958).

The explanations behind the Mills-Reincke phenomenon have remained enigmatic (Cairncross 2003). Contemporaries (Sedgwick and Macnutt 1908) considered two alternatives: (a) that the "other" category would ultimately also be waterborne or (b) that clean water supply somehow enhanced vitality—it was known that after convalescence, typhoid fever patients suffered greatly from secondary infections (Ferrie and Troesken 2004, as cited in Cutler and Miller 2005). Evidence to support the first explanation still has not been found. The results derived from Fewtrell and others (2007) are in line with the second explanation and bring together current competing nutrition and infection-based models (Cutler and Miller 2005) for explaining the dramatically lowered mortality rates in the developed countries over the past century. Recent research has revealed that unintentional consequences of improved hygienic conditions include a steep decline in stomach cancer (Kivi and Tindberg 2006; Parkin 2006) and potential for decreased susceptibility to coronary heart disease (Eriksson 2005).

Conservative Estimates

The Mills-Reincke ratios (the ratio of consequences of malnutrition-diarrhea deaths in subregional WHO estimates) are plotted on a graph with decreasing level of development (using increased risk of diarrheal deaths as a proxy), reflecting the close correlation between diarrheal disease and malnutrition (see figure 4.3). These ratios are small (ranging from 1.5 to 1.9) when compared to those conservatively derived (3.0) for the United States during the early 20th century. One explanation for the relatively more conservative ratios might be current child survival interventions such as measles vaccinations. Although current methodologies to estimate burdens attributable to environmental health use the best available information, the impacts of environmental risk factors are considerably underestimated for several reasons.

The health impacts from environmental risk factors are potentially underestimated in several areas. The first is in the area of perinatal deaths. The 11 percent of these deaths considered to be attributable to the environment (in developing countries) is likely to be a substantial underestimate (Prüss-Üstün and Corvalán 2006). Breman, Alilio, and Mills (2004) suggest that as many as about 60 percent of perinatal deaths could be attributable to rampant malaria in Sub-Saharan Africa.[5] Second, growing evidence (Jefferson and others 2007) shows that ALRIs can be prevented by improving personal and household practices,[6] implying that the WSH-attributable health burden becomes substantially larger (Cairncross and Valdmanis 2006). For example, a randomized clinical trial in Karachi, Pakistan, showed that hand washing reduces ALRIs by 50 percent (Luby and others 2005). Third, a substantial proportion of ALRI deaths associated with indoor air pollution would not have occurred had the children not been malnourished (Ezzati, Vander Hoorn, and others 2004).[7] However, only some of these deaths are included in the consequences of malnutrition. Finally, malaria deaths that are attributable to inadequate mosquito habitat control (for example, 42 percent in Sub-Saharan Africa) do not include the fraction that can be effectively prevented by other environmental management measures, such as indoor residual spraying and bednets.

Similarly, there are several indications that the malnutrition-mediated effects of environmental risk factors ("consequences of malnutrition") are underestimated. First, the consequences of malnutrition in Fewtrell and others (2007) consider only attributable fractions of moderate and severe maternal and childhood underweight—less than or equal to 2 standard deviations (SD)—but do not include the consequences of mild malnutrition (from −2SD to −1SD). Mild malnutrition, the health burden of which is included in other estimates, is the most prevalent form of malnutrition and increases death risk almost twofold (Caulfield and others 2004; Fishman and others 2004). Second, 42 percent of malaria deaths in Sub-Saharan Africa are caused by poor WRM. Of the remaining

58 percent of malaria deaths caused from other factors, however, at least a portion should be attributable to environmental factors through the effect of diarrheal infections on malnutrition and increased malaria fatality rates (see figure 4.1). They are not included because some of these deaths would overlap with those directly attributable to inadequate habitat control. Third, given the price and wage inelasticity among the poorest, income spent on environmental health services (such as water) often takes away from food budgets, thereby contributing to malnutrition. In Khartoum, Sudan, for example, residents of one squatter settlement were found to pay as much as 55 percent of their income to purchase water (Cairncross and Kinnear 1992)—sacrificed at the cost of their food budgets.

A final point on why these current estimates are conservative involves the longer-term perspective. The Barker or fetal-programming hypothesis, mentioned in box 2.2, postulates that malnourishment *in utero* or very early in infanthood makes adults more susceptible to chronic diseases (like coronary heart disease) later in life (Eriksson 2005). This hypothesis means that current poor environmental health conditions in early childhood will result in higher future health effects in terms of chronic diseases (Eriksson 2005).

Areas for Future Research

Given the importance of protecting the fetus from malnutrition and subsequent growth faltering, researchers need to pay more attention to quantifying environmental health problems during pregnancy. This research is particularly important in Sub-Saharan Africa, where the malaria-hookworm association results in maternal anemia and subsequent growth restrictions *in utero*. Understanding the role of environmental factors during pregnancy also has important implications for helping reduce the 4 million perinatal deaths that occur every year.

More clinical trials are needed to test whether ALRIs are water washable, using hand washing or more comprehensive household hygiene interventions (Ahmed and others 1993). Additionally, clinical trials on indoor air pollution are also needed. Studies using deaths as end points would be important when health effects of both WSH and indoor air pollution are established, to overcome current methodological issues that lead to underestimating effects of these factors.

Recognizing the links between infections and malnutrition, future studies should properly account for correlated multiple risk factors. Current studies on the effect of malnutrition on respiratory infections (for example, Fishman and others 2004) do not control for indoor air pollution exposure. Similarly, studies on the effect of indoor air pollution or poor water and sanitation rarely control for malnutrition. In the absence of such controls, the burden from a specific disease (for example, diarrheal infections) attributable to various risk factors (such as WSH or malnutrition) may exceed 100 percent.

Key Messages

- In terms of child health, the environmental health burden is substantially higher than previously estimated if links through malnutrition are incorporated, especially in those subregions, such as Sub-Saharan Africa and South Asia, where both malnutrition and poor environmental conditions coexist.
- In Sub-Saharan Africa, environmental health conditions which cause poorer fetus and child survival, may explain the counterintuitively lower malnutrition rates (when compared with South Asia) from poorer fetus and child survival.
- Even when the environmental health burden is conservatively estimated, a multiplier effect exists for environmental health interventions, where investments addressing environmental risks (for example, lack of water and sanitation) not only reduce diarrhea mortality but also reduce mortality from malnutrition-related diseases and its consequences on educational attainment.
- Future research is needed on environmental health effects during pregnancy, additional disease transmission pathways, and better relative risk estimates to more accurately estimate the environmental health burden in children under five years of age.

Notes

1 The authors' definition was fairly broad and covered 22 conditions, including war, violence, and alcohol.

2 For more details on Comparative Risk Assessment project publications, see http://www.who.int/healthinfo/boddocscra/en/index.html.

3 Fishman and others (2004) cite a recent meta-analysis that shows diarrheal incidence is 23 percent higher (12 to 35 percent, with a 95 percent confidence interval) in children with moderate to severe malnutrition.

4 The distinction between direct and indirect effects of environmental risk factors is conceptual. In practice, when calculating the burden of disease related to these risk factors, the two effects may be difficult to separate.

5 According to a recent meta-analysis of 117 studies by van Geertruyden and others (2004), malaria deaths globally are underestimated by 2 million deaths.

6 This is an alternative complementary explanation to the Mills-Reincke phenomenon (Cairncross 2003) in addition to those provided in box 4.3.

7 All together, about 800,000 deaths annually are associated with indoor air pollution (Smith, Mehta, and Maeusezahl-Feuz 2004).

CHAPTER 5

Estimating the Environmental Health Burden and Costs at the Country Level

IN PART I, new results from the World Health Organization (WHO) on the environmental health burden of disease were presented, and findings were discussed (Fewtrell and others 2007). These new calculations take into account the effects of environmental risk factors (including their effects on malnutrition). In this part, the analysis is taken to the next stage: estimating and translating this burden into economic costs at a country level. With the use of case studies from Ghana and Pakistan, earlier estimates are updated by calculating the effects of environmental risks (including the impacts through malnutrition) and by attempting to estimate the impacts on cognition and learning and on future work productivity and income (see appendix D for a detailed methodology).

Existing Practice in Environmental Health Valuation

The World Bank has made extensive use of valuation techniques to measure the costs of environmental degradation at the country level. Studies of the cost of environmental health effects are mostly based on quantification and valuation of disease-specific mortality and morbidity caused by environmental risk factors. In developing countries, these studies have focused mainly on environmental risks, such as inadequate supply of water, sanitation, and hygiene (WSH); indoor air pollution from household solid fuels; urban (outdoor) air pollution; and sometimes exposure to lead. To value health risks, these analyses have typically

61

looked at different health end points, such as deaths or years of life lost, years lived with disability, incidence of diarrheal and respiratory illness, number of hospital admissions, number of doctor visits, restricted-activity days, number of days of caregiver, and so on.

In general, child mortality has been valued using the human capital approach— that is, the present value of the average individual's income stream—and adult mortality has been valued using both the human capital approach and the value of statistical life, reflecting individuals' willingness to pay for a reduction in the risk of death. Morbidity has been valued by the cost of illness, including the cost of hospital admissions, doctor visits, restricted-activity days, time and income losses from illness, and days of caregiving. A fraction of the average wage is usually taken as the unit value of time lost to illness and caregiver's time (see appendix E for more details).

The cost of environmental health risks at the country level has ranged from 1.2 percent of gross domestic product (GDP) in Tunisia to more than 4.0 percent of GDP in China (see figure 5.1). These costs have typically represented between 40 and 60 percent of total cost of environmental degradation (environmental health and natural resources). Lack of access to water supply and sanitation and poor hygiene often take the largest share of environmental health risks. In Bolivia, El Salvador, Guatemala, and Nigeria, indoor air pollution is particularly important.

Results from these valuation studies—often quoted in various newspapers— have been used to raise awareness on environmental issues in many countries, as well as to attempt to catch the attention of policy-makers. In Algeria, for example, the results were used at the highest political level and proved instrumental in increasing the budget for environmental protection by US$450 million. The studies have also sparked interest in additional work. For example, the governments of Algeria, the Arab Republic of Egypt, and Morocco have commissioned in-depth analyses of the costs of environmental degradation in the water sector. Tunisia is in the process of setting up a unit of environmental economics and policy in the Ministry of Environment to build capacity to undertake such analyses.

Building New Estimates for Environmental Health Costs

Research evidence discussed in chapter 2 highlights the importance of the vicious cycle of malnutrition and infections. This new valuation exercise seeks to estimate previously unaccounted costs associated with how environmental risk factors affect mortality through their effects on malnutrition. In the short term, this calculation means including a portion of the deaths not usually attributed to environmental factors (for example, measles) that would not have occurred if the child had not been malnourished. In the longer term, it means assessing the impact (and subsequent costs) associated with impaired cognition and learning caused by malnutrition (which in turn was affected by environmental factors).

FIGURE 5.1
Cost of Environmental Health Risks

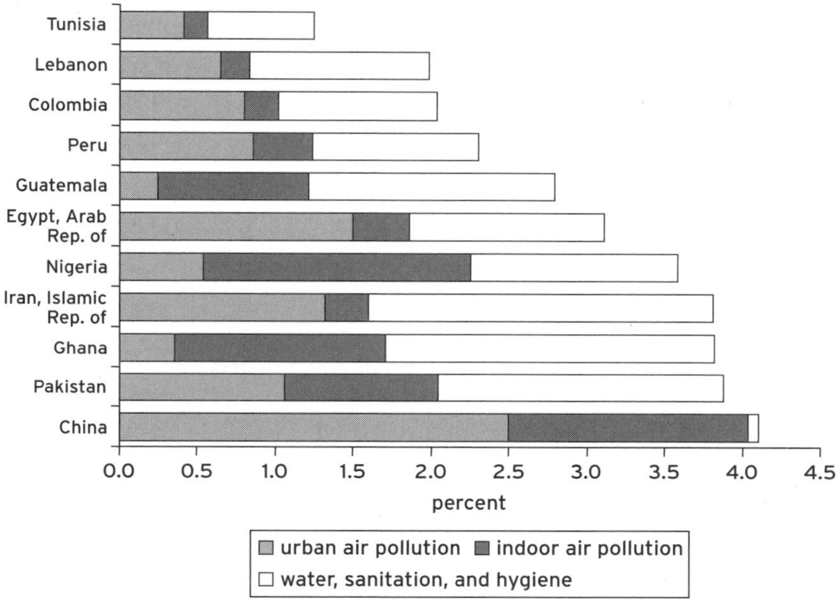

Source: China (World Bank 2007a); Colombia, Ghana, Guatemala, Nigeria, Pakistan, and Peru (Country Environmental Analysis); Arab Republic of Egypt (World Bank 2002), Islamic Republic of Iran (World Bank 2005); Lebanon and Tunisia (Sarraf, Larsen, and Owaygen 2004).
Notes: Economic burden includes the burden from mortality, morbidity, and cost of illness. Morbidity is usually valued using the human capital value approach. Adult mortality is valued by averaging the value of a statistical life approach and the human capital value approach. With the exception of China, the Islamic Republic of Iran, and Nigeria, child mortality is valued using human capital value only. With the exception of China, the value of a statistical life is obtained through benefit transfer of international studies. Water, sanitation, and hygiene mortality is estimated only for children; in China, such mortality excludes lack of sanitation and hygiene costs.

Estimating and Valuing Health Impacts Accounting for Malnutrition

Repeated infections in early childhood contribute to malnutrition. Malnutrition, in turn, makes children more vulnerable to repeated infections. As discussed in chapter 2, numerous studies have found that infections, especially diarrheal disease, in the first few years of children's life impair weight gains, often in the range of 20 to 50 percent (Adair and others 1993; Bairagi and others 1987; Black, Brown, and Becker 1984; Checkley and others 1997; Condon-Paoloni and others 1977; Molbak and others 1997; Rowland, Cole, and Whitehead 1977; Rowland, Rowland, and Cole 1988; Zumrawi, Dimond, and Waterlow 1987).

Valuing Education and Cognition Costs

Children exposed to poor environmental conditions often end up malnourished, which, in turn, impairs their cognitive development and educational performance.

This impairment affects labor productivity, thus reducing individual earnings. Furthermore, malnutrition can lead to delayed primary school enrollment and grade repetition, which, in turn, may affect lifetime income in terms of delayed labor-force entry.

As discussed in chapter 2, a number of recent longitudinal studies (see appendix B) have investigated the effect of early childhood malnutrition on educational outcomes. These studies use data that follow children over time, usually from birth to well into primary or secondary school level, and they contain various anthropometric, schooling, and socioeconomic status indicators. The studies analyzed show that the change in grade attainment ranges from 0.12 years (mild stunting in the Philippines), to 0.91 years (moderate malnutrition in Brazil) and 0.7 years (in Zimbabwe) for each z-score (one standard deviation from the mean). The effect of malnutrition on learning productivity or achievement appears to be even larger, ranging from 0.8 years of grade equivalent in Guatemala and the Philippines for each z-score to 1.8 years for each z-score in severely malnourished children. Delayed primary school entry may vary between two and four months per z-score in the studies reviewed.

Case Studies of Ghana and Pakistan

The previous subsections mention the progress in the methodology for estimating costs associated with environmental health risks—that is, extending the analysis to incorporate the malnutrition-mediated effects of WSH-related infections, as well as estimating the longer-term costs of lower school performance and work productivity from impaired cognitive development and learning capacity. In this section, this adapted methodology is applied to two country case studies: Ghana and Pakistan. These countries were chosen for two reasons: (a) because each belongs to one of the two subregions that have the highest environmental health burden of disease in children under five and (b) because in each country the methodology estimating the effects of environmental factors had been applied earlier, providing an opportunity to compare the "old" (excluding malnutrition) and "new" (including malnutrition) estimates.

In Ghana (see box 5.1)—a country of 22 million people—one in every nine children dies before reaching age five. Under-five mortality in the rural areas is 118 per 1,000 live births as compared with 93 for the urban areas. The country has an inadequate supply of piped water in terms of both population coverage rates and quality of services. On average, access to improved sanitation is also very low at 18 percent (World Bank 2006b), while in the largest cities about 30 percent of the population has household latrines. About 95 percent of households use solid fuels for cooking in Ghana (Ghana Statistical Service 2003). Results from household surveys reveal that among children under five years of age, 10 percent

BOX 5.1
Basic Indicators for Ghana and Pakistan

Ghana

Population: 22.1 million
Children under five years of age: 3.1 million
GDP: US$10.7 billion
Prevalence of malnutrition in children under five: 55 percent[a]
Percentage of population having access to an improved water source:
75 percent
Percentage of population having access to improved sanitation: 18 percent
Energy from biomass products and waste (percentage of total): 69.1 percent
Under-five mortality (per 1,000 live births): 112
Acute respiratory infection prevalence in children under five: 10 percent[b]
Diarrhea prevalence in children under five: 17.9 percent[b]
Malaria incidence per 1,000 (2003): 169.81[b]

Pakistan

Population: 155.8 million
Children under five years of age: 19.0 million
GDP: US$110.7 billion
Prevalence of malnutrition in children under five: 70 percent[a]
Percentage of population having access to an improved water source:
91 percent
Percentage of population having access to improved sanitation: 59 percent
Energy from biomass products and waste (percentage of total): 35.6 percent
Under-five mortality (per 1,000 live births): 99
Acute respiratory infection prevalence in children under five: 24 percent[b]
Diarrhea prevalence in children under five: 26 percent[b]
Malaria incidence per 1,000 (2003): 0.80[b]

Sources: Population figures for children under five are from UNICEF country data; malaria
incidence figures are from RBM (2005); other information is from World Bank (2007h).
a. Percentage of mildly, moderately, and severely underweight children under five years of age.
b. Two-week (14 days) prevalence.

were reported to have had symptoms of acute respiratory illness, and 18 percent
had diarrhea in the two weeks preceding the survey (Ghana Statistical Service
2003). Vectorborne or water-based diseases, such as malaria and schistosomiasis,
are rampant because of inadequate environmental management programs, although
some success has occurred in the increased use of bednets. Ghana is the only Sub-
Saharan country to have dramatically improved food security, according to its
international pledges (FAO 2006).

Pakistan is a country with 156 million inhabitants (see box 5.1). In Pakistan, population growth is coupled with trends in industrialization and urbanization. The under-five mortality rate for Pakistan is estimated to be 99 per 1,000 live births. Both rural and urban areas suffer from underinvestment in water supply and sanitation. Sewerage systems, where they exist, are usually poorly maintained and often contribute to the cross-contamination of water systems and the groundwater. In rural areas, more than 40 million people lack access to safe drinking water, while 60 million lack adequate sanitation. Inadequate quantity and quality of potable water and poor sanitation facilities and hygiene practices are associated with a host of illnesses, such as diarrhea, typhoid, intestinal worms, and hepatitis. About 70 percent of households use solid fuels for cooking in Pakistan (Federal Bureau of Statistics 2005). Dehydration attributable to diarrhea is a major cause of mortality among children in Pakistan. In 2001 to 2002, some 12 percent of children under five were reported to have suffered from diarrhea in the previous 30 days (Krupnick, Larsen, and Strukova 2006).

Results for Ghana and Pakistan

This part of the chapter presents and discusses the new estimates for the economic costs attributed to environmental risk factors in Ghana and Pakistan. The first subsection presents the results of the costs from environmental risk factors (excluding malnutrition), alluding also to estimates provided in previous studies. The next subsection presents the context and results for the total effects—including malnutrition—of these environmental risk factors. The final subsection presents estimates for the economic costs of poor school performance and lost work productivity that are attributed to impaired cognitive development and learning from the infection-malnutrition links.[1]

The Burden of Environmental Risk Factors If No Change in Malnutrition

Earlier studies have reported the cost of environmental health effects in Ghana (Larsen 2006) and the cost of environmental degradation in Pakistan (Krupnick, Larsen, and Strukova 2006). In this study, the same methodology is applied to recently published baseline data to estimate mortality from indoor air pollution and inadequate water supply, sanitation, and hygiene. This report uses WHO country estimates of cause-specific mortality for 2002, adjusted to reflect under-five child mortality rates in 2005. Solid fuel use prevalence is also updated from the *Pakistan Social and Living Standards Measurement Survey 2004–05* (Federal Bureau of Statistics 2005).

Estimated environmentally attributable fractions of child mortality in Ghana and Pakistan are presented in table 5.1. These fractions are estimated as follows. The attributable fraction for acute lower respiratory infection (ALRI) is related

TABLE 5.1
Environmentally Attributable Fractions of Child Mortality, Keeping Malnutrition Unchanged

Disease	Ghana (%)	Pakistan (%)
Pneumonia/ALRI	55.2	47.8
Diarrhea	88.0	88.0
Malaria	42.0[a]	40.0

Sources: Compiled by World Bank team on the basis of Federal Bureau of Statistics 2005; Ghana Statistical Service 2003; Prüss-Üstün and Corvalán 2006.
a. In the case of Ghana, the subregional environmentally attributable fraction for malaria of 42 percent may be an overestimate because of mosquito ecology, which demands that environmental manipulation be supplemented by bednets and indoor residual spraying.

TABLE 5.2
Estimated Mortality in Under-Five Children from Environmental Risk Factors, 2005

Disease	Ghana		Pakistan	
	Under-5 Child Mortality	Mortality from Environmental Risk Factors (Excluding Malnutrition)	Under-5 Child Mortality	Mortality from Environmental Risk Factors (Excluding Malnutrition)
Pneumonia/ALRI	11,800	6,508	97,100	46,435
Diarrhea	9,900	8,712	96,200	84,656
Malaria	22,600	9,492	1,300	520
Total	75,800[a]	24,712	402,600[a]	131,611

Source: Compiled by World Bank team on the basis of WHO 2004.
a. Includes all under-five child mortality, not just mortality from ALRI, diarrhea, and malaria.

to indoor air pollution and is estimated from the share of households using solid fuels and the relative risk of ALRI in children from such fuel use—that is, relative risk equals 2.3—as reported by Desai, Mehta, and Smith (2004). About 95 percent of households use solid fuels for cooking in Ghana (Ghana Statistical Service 2003), and 70 percent use such fuels in Pakistan (Federal Bureau of Statistics 2005). The attributable fraction for diarrhea is related to WSH and is based on regional estimates by WHO. The attributable fraction for malaria is related to water resource management and is from Prüss-Üstün and Corvalán (2006).

Under-five child mortality in 2005 was 112 and 99 per 1,000 live births in Ghana and Pakistan, respectively (World Bank 2007h). Those figures imply about 76,000 annual deaths in Ghana and about 400,000 in Pakistan.[2] Accordingly, the mortality attributed to different diseases (cause-specific mortality) is estimated from WHO (2004). The results show that when malnutrition is not fully accounted for, environment-related mortality is estimated at about 25,000 children per year in Ghana and 132,000 in Pakistan, or about one-third of all child mortality in each country (table 5.2).

The Burden of Environmental Risk Factors and Malnutrition

This subsection presents the revised estimates in a country context for assessing the effect on mortality relating to environmental risk factors, such as poor WSH, indoor air pollution, inadequate water resource management, and malnutrition. First, the subsection presents child malnutrition status in the two countries to provide a sense of the prevalence of stunting and underweight in children (see table 5.3). Second, it examines the relationship between diarrheal infections and malnutrition status and estimates the malnutrition rates in the absence of infections. Third, the subsection estimates the impact of environmental factors—including the effect of infections on malnutrition—on health outcomes in children under five (mortality). Finally, a separate subsection addresses the effect of infection-malnutrition links on cognition and learning, as well as the accompanying effects on education and costs of lifetime income losses.

Malnutrition in Ghana and Pakistan. Malnutrition prevalence is high in many developing countries. In Ghana, about 55 percent of children under the age of five years were mildly, moderately, or severely underweight in 2003,[3] which is still 15 percentage points less than the average in Sub-Saharan Africa (Fishman and others 2004). About 70 percent of children under five in Pakistan were mildly, moderately, or severely underweight in 1991, with minimal change in 2001 to 2002.[4]

An important feature of malnutrition rates in developing countries such as Ghana and Pakistan is that it tends to rise very quickly in the first stages of life and then levels off or declines. More than 90 percent of the children in the *Ghana Demographic and Health Survey* (Ghana Statistical Service 2003) were not underweight in their first three months of life; moderate and severe underweight was

TABLE 5.3
Malnutrition Rates in Children under the Age of Five

	Malnutrition Rate (%)	
Indicator	Ghana	Pakistan
Underweight (weight-for-age)		
Mild (–1 to –2 SD)	33	29
Moderate (–2 to –3 SD)	17	27 (26)
Severe (less than –3 SD)	5	14 (12)
Total (less than –1 SD)	55	70
Stunting (height-for-age)		
Mild (–1 to –2 SD)	28	22
Moderate (–2 to –3 SD)	19	20 (19)
Severe (less than –3 SD)	11	30 (18)
Total (less than –1 SD)	58	72

Sources: Compiled by World Bank team on the basis of data from Ghana Statistical Service 2003; National Institute of Population Studies 1992. Figures in parentheses are based on PIDE 2003.

practically nonexistent in this age group. The prevalence of moderate and severe underweight peaked at over 35 percent in children 10 to 13 months of age, however, and then gradually declined to the age of five years. The prevalence of mild underweight reached about 40 percent in children age 18 to 21 months and then leveled off to the age of five years. A similar evolution can be inferred from Pakistani data. More than 97 percent of the children were not underweight in their first month of life, but by 9 to 10 months of age, the prevalence of moderate and severe underweight rose rapidly to 40 percent, stabilizing to a mean of 45 percent for children age 18 to 59 months.

In global comparisons, too, children 1 to 3 months of age in Ghana and Pakistan have similar weight-for-age scores (indicator for underweight) to the international reference population, but the scores fall considerably below the international reference during the first year of life (see figure 5.2). By the age of 10 to 15 months, children from both countries are significantly underweight, with more than 70 percent malnourished.

Health effects of malnutrition. The malnutrition-attributable fractions for different diseases are calculated for Ghana and Pakistan (see table 5.4) using estimates of increased risk of cause-specific mortality and all-cause mortality in children under five years of age with varying levels of malnutrition. The results indicate that, in Ghana, 44 to 63 percent of under-five mortality from pneumonia-ALRI, diarrhea, measles, malaria, and other causes is due to malnutrition. The corresponding attributable fractions for Pakistan are in the range of 50 to 68 percent because of higher underweight malnutrition rates in Pakistan. In addition to these malnutrition-related mortalities, Fishman and others (2004) include 100 percent of protein-energy malnutrition mortality and morbidity and a share of mortality from perinatal conditions (low birth weight) associated with mothers' low pre-pregnancy body mass index (less than 20 kilograms per square meter). About 16 percent of infants had low birth weight (less than 2,500 grams) in Ghana in 2003, as did about 19 percent of infants in Pakistan in the past decade (UNICEF 2005).

Infections and malnutrition. Household survey data from Ghana and Pakistan provide some insights into the close links between diarrheal prevalence and underweight in children. In Ghana, the two-week diarrheal prevalence was 15 to 18 percent in children under five years and 20 to 22 percent in children under three years, according to the three most recent demographic and household surveys. This diarrheal prevalence differs dramatically by degree of underweight in Ghanaian children age 3 to 23 months (panel a of figure 5.3), with prevalence in severely underweight children age 3 to 11 months reaching 3.5 times higher than that in children who are not underweight. A similar pattern, although with less drastic differences across malnutrition categories, exists in Pakistan. With a two-week diarrheal prevalence of 15 percent in children under five years, the prevalence in

FIGURE 5.2
Weight-for-Age Distribution of Children in Ghana and Pakistan

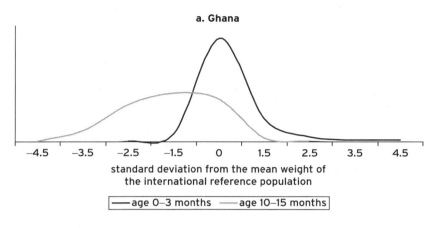

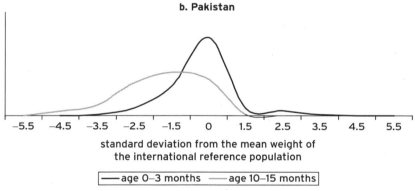

Sources: Compiled by World Bank team on the basis of data from Ghana Statistical Service 2003; National Institute of Population Studies 1992.

TABLE 5.4
Malnutrition-Attributable Fractions of Child Mortality

Cause of death	Ghana (%)	Pakistan (%)
Pneumonia-ALRI	54.1	60.1
Diarrhea	63.2	67.8
Malaria	57.6	63.1
Measles	43.8	50.5
Protein-energy malnutrition	100.0	100.0
Other causes (excluding perinatal conditions)	55.8	61.5

Source: Compiled by World Bank team on the basis of Fishman and others 2004.

FIGURE 5.3
Two-Week Diarrheal Prevalence Rate by Age and Underweight Status in Ghana and Pakistan

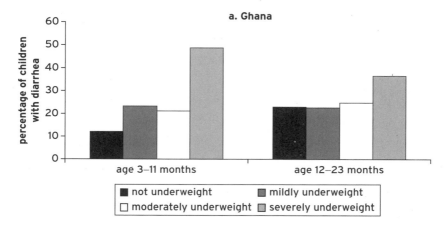

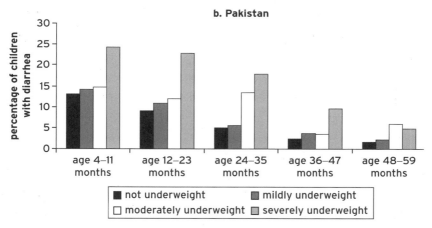

Source: Compiled by World Bank team on the basis of data from Ghana Statistical Service 2003; National Institute of Population Studies 1992.

severely underweight children is 50 percent higher than in children who are not underweight (panel b of figure 5.3).

Because these data are cross-sectional (both indicators measured at the same point in time), no causal relationship between underweight and diarrheal prevalence can be determined. However, the pattern is consistent with research findings in other countries that show that diarrheal infections lead to weight gain retardation in children under five years. This research literature finds that the weight gain retardation often is in the range of 20 to 50 percent. Diarrheal infections are most frequent during ages 6 to 11 months and 12 to 23 months, so weight gain retardation from diarrheal infections can be expected to be greatest in this period; in Ghana and Pakistan, it is particularly great up to the age of 12 months, during which underweight malnutrition increases rapidly.

Malnutrition in the absence of infections: the counterfactual. Results from the research literature are used to estimate the retarding effect that diarrheal infections have on weight gain. This information sheds light on an important question: what would be the counterfactual level of malnutrition in children who were not exposed to diarrheal infections?

Current underweight malnutrition rates in children under five years from the 2003 demographic and health survey in Ghana (Ghana Statistical Service 2003) and the 1991 survey in Pakistan (National Institute of Population Studies 1992), as well as the estimated malnutrition rates in the absence of diarrheal infections that retard weight gain are presented in figure 5.4. Basically, the figures for both Ghana and Pakistan starkly show that in the absence of diarrheal infections, children under five years in both countries would climb up the nutritional status ladder—with practically no severely underweight children in either country and with the number of moderately underweight also substantially reduced.[5]

Calculating the burden of environmental risk factors, including malnutrition-mediated effects. Information from the previous subsections provides the necessary building blocks for estimating the effects of environmental health risk factors. The counterfactual population exposure rates estimated are used to compare the current world with a world without environmental risk factors in order to compute the environmental health risk—that is, the attributable fraction (table 5.5). Appendix D (equation D.2) is used. In making these estimates, the study takes a conservative approach. In fact, the parameters on which attributable fractions are based are likely to be affected by omitted variable bias (see appendix E). To deal with this problem, the study attempted to correct such parameters by using formulas from econometrics literature. The methodology is graphically represented in figure 5.5. The indirect effect of environmental health risks is equivalent to the difference in mortality between the malnutrition-attributable fractions of different diseases (right bar) and the consequences of malnutrition in the absence of diarrheal infections (center bar). Notice that the joint direct and indirect effects of environmental health risks (left bar) are not necessarily equal to the sum of the two effects computed separately. This problem is common in the presence of risk factors that interact with each other (see box 5.2). Figure 5.5 shows that additional pneumonia-ALRI, diarrhea, malaria, measles, protein-energy malnutrition, and other deaths represented in the light-gray area have not been accounted for previously.

The results (shown in table 5.5) demonstrate that when both direct and indirect effects from diarrheal infections on malnutrition are included, the mortality

FIGURE 5.4
Underweight Malnutrition Rates in Children with and without Diarrheal Infections in Ghana and Pakistan

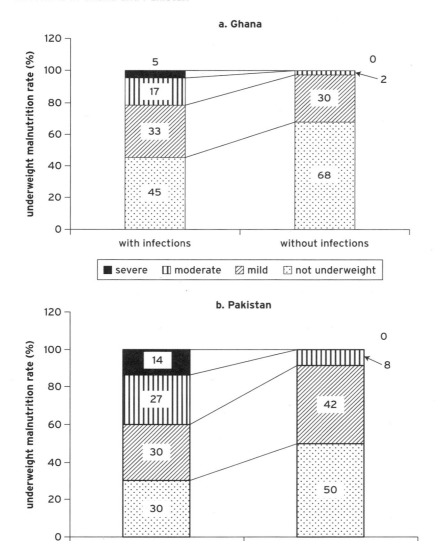

Sources: Compiled by World Bank team on the basis of data from Ghana Statistical Service 2003; National Institute of Population Studies 1992.

TABLE 5.5
Environmentally Attributable Fractions and Child Mortality with Malnutrition-Mediated Effects

Disease	Ghana Revised Effects (Number of Deaths)	Ghana Attributable Fraction (%)	Pakistan Revised Effects (Number of Deaths)	Pakistan Attributable Fraction (%)
Pneumonia-ALRI	7,375	62.5	58,260	60.0
Diarrhea	9,009	91.0	88,504	92.0
Malaria	14,035	62.1	826	63.5
Measles	58	28.8	8,316	30.8
Protein-energy malnutrition	400	40.0	1,359	45.3
Other causes (excluding perinatal conditions)	4,826	38.3	30,164	39.9
Total	35,702	47.1	187,429	46.6

Source: Compiled by World Bank team.

FIGURE 5.5
Calculating Revised Estimates (Indirect and Direct Effects)

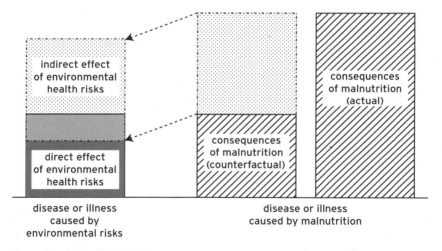

Source: Compiled by World Bank team.

from environmental factors is estimated at 36,000 in Ghana and 187,000 in Pakistan annually, or about 47 percent of all child mortality in both countries.

Translating those deaths into economic costs will help quantify the magnitude of environmental health risks in real terms. In this analysis, the human capital approach (a conservative approach) is used to value this mortality. For Ghana, the earlier estimates of costs of environmental factors add up to ₵370 million (US$412 million), which is equivalent to 3.8 percent of the country's GDP.[6]

BOX 5.2

Attributable Fractions and Burden of Disease When Multiple Risk Factors Are Present

The concept of an attributable fraction (AF) is the essential building block of the burden of disease methodology. An AF tells what proportion of a disease is caused by any given risk factor or combination of risk factors. To take into account the total effects of environmental risks on child health—including those effects mediated through malnutrition—one needs to allow for the simultaneous presence of multiple risk factors. Whether one or more risk factors are being controlled should be clearly stated when one calculates AFs.

Tables 5.1, 5.4, and 5.5 present the fraction of deaths that are in turn attributed to (a) environmental health risks alone (table 5.1), (b) malnutrition alone (table 5.4), and (c) all relevant environmental health risks, including their effect through malnutrition (table 5.5). This calculation is made for specific causes of death (for example, ALRI, diarrhea, malaria, measles, and protein-energy malnutrition). AFs in tables 5.1 and 5.4 measure the cases of death from a disease that would be eliminated by removing a risk factor from the population while leaving the exposure to other risk factors unchanged.

Take Ghana as an example. The fraction of ALRI attributable to indoor air pollution is 55.2 percent (table 5.1), and the fraction of ALRI attributable to malnutrition is 54.1 percent (table 5.4). One could also compute the fraction of a disease that could be avoided if one eliminated both risk factors. In the case of Ghana, eliminating both indoor air pollution and malnutrition would reduce ALRI by nearly 80 percent. Notice that the sum of AFs calculated for individual risk factors can be more than 100 percent and can differ considerably from the AF calculated for the risk factors jointly. This result is common, as shown in Chatterjee and Hartge (2003).

This study controls for risk factors jointly (for example, environmental health risks and malnutrition) but allows malnutrition to be reduced only to the level that would be present without diarrheal infections. It answers this question: by how much would a disease (for example, ALRI) be reduced if there were no environmental health risk (for example, indoor air pollution) and if the burden of ALRIs that was caused by lack of water, sanitation, and hygiene and was not related to indoor air pollution were avoided? A problem one faces is the difficulty in saying what fraction of the resulting AF is attributable to each risk factor.

The "new" (including malnutrition) costs amount to ₵540 million (or US$595 million), which is equivalent to 5.6 percent of Ghana's GDP—a 47 percent increase over previous estimates.[7] In Pakistan, the earlier estimated costs of environmental factors amount to PRs 195 billion (US$3.2 billion), which is equivalent to 2.9 percent of GDP. The "new" (including malnutrition) costs raise that figure to 4.1 percent of the country's GDP—a 41 percent increase over previous estimates.

The lower cost in Pakistan results from its lower child mortality rate and its lower birth rate, which gives a lower share of children under five in the total population.

Impact on Education: Affecting School Performance and Future Work Productivity

Chapter 2 discussed another indirect effect of the link between repeated infections in early childhood and malnutrition: the impairment of cognitive development, resulting in poor school performance and, subsequently, in a loss of lifetime earnings. Many studies have documented the effect of early childhood malnutrition on cognitive function, school enrollment, grade repetition, school dropout rates, grade attainment, and test scores among school-age children (see appendix B). The majority of these studies use stunting (low height for age) as a measure of malnutrition, which is considered an indicator of chronic nutritional deficit. Here, for the first time, this study attempts to estimate the loss in lifetime earnings from stunting in Ghana and Pakistan and the share of this loss that could be avoided if children were not exposed to repeated diarrheal infections in early childhood.

Effects of malnutrition on education outcomes applied to Ghana and Pakistan for estimation of the cost of malnutrition are presented in table 5.6, stratified by height-for-age deficit. The magnitude of these effects is based on the review of longitudinal studies. Effects analyzed include those on (a) grade attainment, (b) school achievement (learning productivity) in terms of grade equivalents, (c) delayed primary school enrollment, and (d) grade repetition. The latter two effects result in delayed labor-force entry.

Psacharopoulos and Patrinos (2004) provide a review of studies of returns to investment in education. Private returns are estimated on the basis of private cost of schooling and lifetime income. Social returns include social cost of providing

TABLE 5.6
Effects of Malnutrition on Education

Effects of Stunting	Height-for-age z-Score		
	–1 to –2	–3 to –2	Lower than –3
Loss in grade attainment (school years)	0.50	1.0	1.5
Loss in school achievement (grade equivalents)	0.8	1.6	3.6
Total grade equivalent losses (grade attainment and achievement)	1.30	2.60	5.10
Delayed enrollment (years)	0.4	0.8	1.2
Grade repetition (percentage of students with one repetition)	2	5	9
Delayed labor-force entry (years) caused by delayed enrollment and grade repetition	0.42	0.85	1.29

Sources: Compiled by World Bank team on the basis of review of longitudinal studies. For more details see appendix B.
Note: Stunting is defined as height-for-age z-score lower than –1.

schooling, which is most often significantly higher than private cost because of subsidized provision of education. Social returns are therefore lower than private returns to investment in education.[8]

Returns to one additional year of schooling are calculated estimates from Psacharopoulos and Patrinos (2004). On average, the return to one additional year of schooling is 9.7 percent. The rate of return is 11.7 percent in Sub-Saharan Africa and 9.9 percent in Asia. These rates of return are applied to Ghana and Pakistan, respectively, to estimate the cost of impacts of malnutrition on education outcomes.

To estimate the lifetime income for children in Ghana and Pakistan, the analysis considered existing income differentials between children in different nutritional status (from not malnourished to severely malnourished). This differentiation is important because malnourished children live disproportionately in poorer households; given that poverty is correlated with lower socioeconomic status, such as lower educational attainment of parents, it is, in turn, correlated with lower educational attainment of their children. Malnourished children are more likely to come from lower-income households than are children who are not malnourished. Therefore, the income potential of malnourished children is affected not only by the action of malnutrition on educational performance but also by the socioeconomic status of the household they grew up in (see appendix D for methodology on calculations).[9]

About 630,000 children are born annually in Ghana.[10] Thirty-five percent of these children are moderately or severely stunted by the time they reach one to two years of age, and an additional 30 percent are mildly stunted (height-for-age z-score –1 to –2). On the basis of these prevalence rates, the effect of stunting on educational outcomes, income losses from loss in education, and a labor-force participation rate of 75 percent, the annual cost of stunting that is attributable to early childhood diarrheal infections is estimated to be about 3.8 percent of GDP.

About 3.66 million children are born annually in Pakistan.[11] According to the Pakistan *National Nutrition Survey 2001–2002* (PIDE 2003), prevalence of moderate and severe stunting in children under five years was 19 and 18 percent, respectively. No prevalence rate is reported for mild stunting. A rate of 25 percent is therefore assumed, in accordance with cross-country comparison. On the basis of these prevalence rates, the effect of stunting on educational outcomes, income losses from educational losses, and a labor-force participation rate of 60 percent, the annual cost of stunting attributable to early childhood diarrheal infections is estimated to be about 4.7 percent of GDP (see appendix D for methodology on calculations).

Conclusion

Malnutrition-mediated effects of environmental risk factors are estimated to have a considerable social and economic burden in Ghana and Pakistan (see figure 5.6

FIGURE 5.6
Final Results of Ghana and Pakistan Case Studies

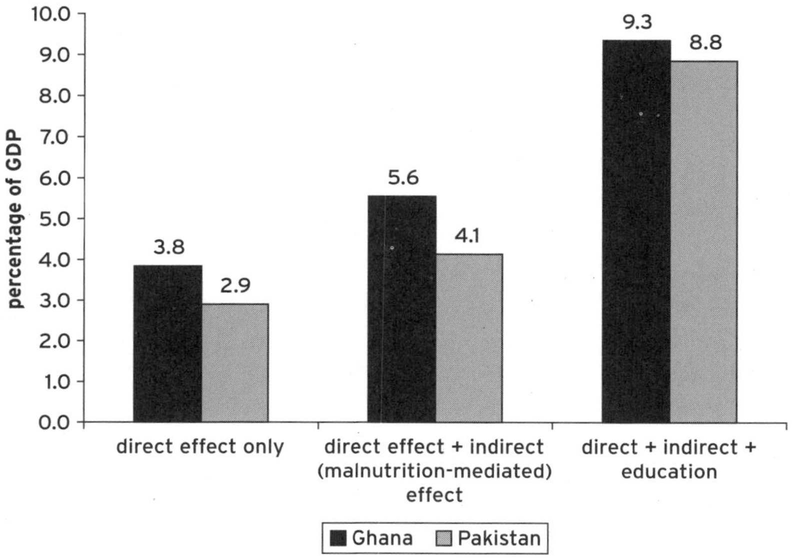

Source: Compiled by World Bank team.

and table 5.7). For Ghana, the cost of total health effects (including malnutrition) amounts to ₵537 million (US$595 million). When longer-term effects on education and income are included, this figure may reach as high as ₵894 million (US$1 billion)—equivalent to 9.3 percent of GDP. Similarly, for Pakistan, the cost of total health effects from environmental risk factors amounts to PRs 278 billion (US$4.6 billion). And with longer-term effects on education and income, this figure may reach as high as PRs 595 billion (US$9.9 billion), or equivalent to 8.8 percent of the country's GDP. Put in perspective, the costs of these total effects are nearly three times higher than the cost of the effect of environmental risk factors only (excluding malnutrition).

The issue of temporality is important to recognize while analyzing these results. These figures indicate that the health benefits from reducing environmental risks are high even if such benefits will accrue far into the future. With the focus on children under five years of age, environmental health interventions implemented today to avoid deaths and illness and to increase education attainment will generate returns only a few years from now. Conventional decision making in developing countries often lacks this longer-term perspective when policy-makers choose appropriate interventions to benefit child health. By providing a better understanding of the links between environmental risks (the causes) and child health

TABLE 5.7
Annual Cost of Direct and Indirect Impact of Environmental Risk Factors in 2005

	Ghana				Pakistan			
	Annual Deaths	Cost (₵ million)	Cost (US$ million)	Cost (% of GDP)	Annual Deaths	Cost (PRs billion)	Cost (US$ million)	Cost (% of GDP)
Estimation excluding malnutrition-mediated effects								
Mortality effects	24,712	371	412	3.84	131,611	195	3,250	2.90
Estimation including malnutrition-mediated effects								
Mortality effects	35,702	537	595	5.55	187,429	278	4,633	4.13
Education effects		367	407	3.79		317	5,281	4.71
Total effects		904	1,002	9.34		595	9,914	8.84

Source: Compiled by World Bank team.
Note: ₵ = Ghanaian new cedi.

(the effect), as well as the timing of these effects, this study hopes to influence the way that governments choose between short-term and long-term interventions.

Given the importance of time in these results, the methodology relating to discounting becomes very relevant. In this report, the calculations used a 3 percent discount rate, which is consistent with the rate used in WHO burden of disease calculations. The purpose of discounting is to recognize a general preference for benefits that happen earlier. Different discount rates are used in different situations. When one looks at public sector decisions, a social discount rate is preferred to a private one (that is, the rate a commercial bank would apply on a loan). Pearce and Ulph (1999) report an estimated social rate of return on investments for industrial countries between 2 and 4 percent. In appendix D, a sensitivity analysis applied to Ghana varies the discount rate from 3 percent to 5 percent (a more conservative discount rate) to show how the results would vary. In the case of Ghana, a 5 percent discount rate would reduce the total cost estimates from 9.2 percent to 4.6 percent of GDP, and in Pakistan it would reduce them from 8.6 percent to 4.3 percent of GDP. The considerable difference is explained by the fact that a large part of the benefits from interventions (for example, higher incomes owing to education) will happen several years from now. Box 5.3 provides guidelines for policy-makers in interpreting the results.

Next Steps

For policy-makers to prioritize among various environmental health interventions, further analyses and studies are required to guide their decisions along two

---

BOX 5.3
How Policy-makers Should Interpret These Results

Using Ghana and Pakistan as examples, this study estimates the annual costs associated with environmental risks for children under five years of age. These estimations, carried out at a national level, provide a sense of the magnitude of these costs (about 9 percent of a country's GDP) to policy-makers and highlight the urgent need to position environmental health at the center of all child survival strategies. These figures help carry the message that environmental risks to child health are considerable and that such risks impose a significant economic burden.

That being said, these figures do *not* help policy-makers prioritize (in terms of costs and capacity to implement) among the various types of environmental health interventions that are available to them. For example, if policy-makers in Pakistan choose to address diarrhea in children, should they invest in sanitation, in hygiene promotion, in increased water availability, or in some combination of these interventions? These choices, as well as the underlying framework that helps determine their priority, are important because they help policy-makers translate the environmental health "agenda" into concrete programs and actions addressing environmental health.

---

major paths: one relating to *costs and benefits of specific interventions*, which has implications for budgetary and resource allocation decisions, and the second relating to *institutional capacity*, which helps ascertain the required roles and responsibilities of various agencies in implementing these interventions. These analyses and studies need to be carried out separately for specific countries because the priorities may differ according to the country's current level of development, its population health profile and the relative exposures to different disease, and its capacity to implement specific interventions.

In terms of economic approaches, two main methods are typically used: cost-effectiveness analysis and cost-benefit analysis. Cost-effectiveness analysis allows researchers to compare policy options that provide outcomes measurable with a common metric (for example, per child life saved). An example would be an assessment of how costs of providing latrines to a village compare with costs of providing access to safe drinking water per child life saved. Cost-benefit analysis, in contrast, can be applied to compare options that do not necessarily provide outcomes measured using the same metric by reducing the benefits of each option to a common numeraire (that is, by assigning a monetary value for the various policy options).

The choice of method used to carry out an economic analysis at a country level must be made after examining the availability of relevant data. For Ghana and Pakistan, for example, this choice means looking at specific environmental risks, identifying possible interventions that would address them, and then assessing how much data on the costs and benefits of these interventions are available. In other country-level studies, cost-benefit analyses have helped governments prioritize among available interventions. In Peru, for example, the World Bank's country environmental analysis reported that drinking-water disinfection and hand-washing programs had the highest cost-benefit ratio for addressing water- and sanitation-related risks, while shifting households from unimproved to improved cookstoves was the best recommendation for interventions to address indoor air pollution (World Bank 2007g).

. Policy-makers in developing countries also need to choose programs to address environmental risks in accordance with their country's institutional capacity to implement these interventions. The next chapter provides some discussion of the generic roles that national and local governments can play in delivering and managing environmental health interventions. A more in-depth country-level institutional analysis is required about what coordination mechanisms exist between ministries and how mandates and budgets are assigned. Such an institutional analysis would provide guidance on how to better define the roles and responsibilities of different agencies related to addressing public environmental health services and how to correct incentive structures to raise the profile of the environmental health agenda.

Key Messages

- When malnutrition-mediated health effects attributed to environmental risk factors are included, the total costs of the environmental health burden are at least 40 percent higher for both Ghana and Pakistan. When longer-term effects of malnutrition (partly attributed to environment-related infections) on education and income are considered, the estimated annual cost may be as high as 9 percent of a country's GDP. This social and economic burden is not trivial, and it highlights the urgent need for policy-makers to position environmental health at the center of all child survival strategies.

- Next steps to translate these messages into concrete actions include carrying out cost-effectiveness and cost-benefit analyses at a country level to guide decision-makers in prioritizing among the various available interventions. An in-depth country-level institutional analysis would also be needed to assess the capacity of the country to implement these interventions.

Notes

1 Details on the steps in the methodology to calculate these costs for Ghana can be found in appendix D.

2 These estimates of mortality are based on crude birth rates and population in 2005.

3 The *Ghana Demographic and Health Survey 2003* (Ghana Statistical Service 2003) does not provide figures for mild underweight (−1 to −2 standard deviation). This figure was therefore calculated from the survey's household data.

4 Underweight malnutrition rates in Pakistan were practically the same in 1991 (National Institute of Population Studies 1992) and 2001 to 2002 (PIDE 2003). The original survey data from the nutrition survey (PIDE 2003) were not readily available for this study, so data from the *Pakistan Demographic and Health Survey 1990/91* (National Institute of Population Studies 1992) were used instead. The survey for 2006 was being compiled at the time of this study. The 2006 survey may show malnutrition rates and diarrheal disease that differ from some of the patterns found in the earlier survey.

5 This finding is based on the fact that 35 percent of children's weight deficit, relative to the international reference population, is from infections, which is well within the range found in research studies.

6 Ghanaian currency is denominated in new cedi ($\mathbb{C}$). The Ghanaian new cedi—U.S. dollar exchange rate is from 2007 (US\$1 = $\mathbb{C}$9,375).

7 A working life from 15 to 65 years is applied. Annual income is approximated by GDP per capita. Income is discounted at 3 percent. Real income growth is 2 percent per year.

8 Social returns do not (in almost all studies) include possible external benefits of education.

9 Income is discounted at a social rate of time preference of 3 percent per year. Social cost of providing a year of schooling is assumed to be 25 percent to 50 percent of average annual income per person.

10 By the time the children turn five years old, 11 percent of them will have died. Under-five child mortality rate in Ghana is 112 per 1,000 live births in 2005 according to World Bank (2007h).

11 By the time the children turn five years old, 10 percent of them will have died. Under-five child mortality rate in Pakistan is 99 per 1,000 live births in 2005 according to World Bank (2007h).

PART III

Experiences

CHAPTER 6

Approaches to Environmental Health

THE PRECEDING CHAPTERS PROVIDE EVIDENCE of the importance of environmental health interventions in addressing child health. Chapter 2 explained the science and the epidemiological underpinnings of the malnutrition-infection cycle and its implications for a child's health both *in utero* and in early childhood. Chapter 3 illustrated how environmental health interventions can appropriately complement and supplement other health and environmental programs that address child health. Chapter 4 then went on to demonstrate the scale of the environmental health burden in children under five and how that burden is especially important in some subregions with high levels of malnutrition and poor environmental conditions. Using two country examples—Ghana and Pakistan—chapter 5 then analyzed the consequent economic burden that these environmental health risks pose.

Clearly, the issue of environmental health is critical for child health—with the "why," "to what extent," and "where" explained in earlier chapters. What remains is the "how": how can these environmental health issues be addressed successfully in developing countries? This chapter begins with a historical review of environmental health, outlining the trends in the evolution of environmental health functions in developed countries and highlighting how circumstances have led to the relative neglect of environmental health in the development agenda. The chapter provides illustrative examples of how different developing countries have incorporated environmental health activities within other health, nutrition, and infrastructure programs. Identifying some common elements from these examples, the chapter then discusses—through an institutional lens—the roles that national and local governments can play in delivering and managing environmental health interventions.

History of Environmental Health

Environmental health is among the earliest public health activities on record. Through the ages, today's developed countries have invested in environmental health interventions, focusing on sanitation, cleanliness, and hygiene. The history of environmental health can be loosely divided into four generations—with shifts in the nature and focus of interventions. Looking at the evolution of environmental health functions over the past few centuries helps to understand the context behind today's environmental health agenda in developing countries.

First Generation: Origins of "Miasma" Theory

In Classical Greece, the source of disease was attributed to bad air or "miasmatic odors," and interventions sought to fight disease through cleanliness and hygiene (Rosen 1958). Subsequently, as the Roman Empire conquered the Mediterranean world, it incorporated the ideas of cleanliness and hygiene into infrastructure projects, which included the great sewerage system (the *cloaca maxima*), baths, fresh water supplies, health facilities, suitable latrines, proper ventilation, and the organization of the health sector (Rosen 1958).

For the English sanitary reformers of the mid-19th century, miasma continued to prevail in explanations of disease transmission. In 1854, however, John Snow traced an outbreak of cholera in London to a common drinking-water source, refuting the miasma hypothesis. The 1871 *Report of the Royal Sanitary Commission* listed 11 essential health services, mostly environmental in nature, that covered access to improved water and sanitation, provision of a sewerage system, healthiness of dwellings, and inspection of food (Kotchian 1997; WHO 2007e).

Second Generation: Sanitation, Vaccinations, Multisectoral Interventions

Refuting the dominant theory of miasma, Louis Pasteur, by the late 19th century, had developed the germ theory that proposes that microorganisms are the cause of many diseases. This theory paved the way for more targeted and cost-efficient environmental and public health interventions, such as chlorination of water supplies, pasteurization of milk (see box 6.1), vaccinations, improvements in food safety and vector control, and introduction of a more effective sewerage system (Berg 2005). Many of the newly adopted health protection policies were

---

BOX 6.1
Combating Disease through Improved Milk

In the early 20th century, one avenue of attack on preventable deaths of children was through improvements in milk. In New York, cross-sectoral collaboration between different government and nongovernmental agencies began as they worked together to combat disease. Nurses visited mothers of newborn babies and promoted hygiene and breastfeeding of babies (Moser Jones 2005). Organized milk stations were set up to provide clean and preheated milk to inner-city inhabitants. Inspectors began to scrutinize milk production facilities, farms, and associated workers to ensure hygienic conditions for milk (Moser Jones 2005). Railway companies were advised to have their milk cars properly refrigerated. The city authority began to supervise the distribution and sale of milk.

The New York City Health Department took the lead in coordinating the efforts of its mandate to secure a safe and hygienic food chain, from farm to child. In many cases, it worked closely with the courts to enforce regulations. Today, different mandates in the legal framework perform the same task in the different stages of the production process (Rosen 1958).

---

based on field evidence and the result of multisectoral collaboration. After local health protection activities had proved successful and enough local level experience had been accumulated, the state played a key role in scaling up the interventions (Rosen 1958).

Third Generation: Hands-on Scaling Up

In the beginning of the 20th century, much of environmental health work focused on providing rural sanitation (Berg 2005). This period was characterized by a community-based approach in which public health nurses made visits to homes and then informed sanitarians, who would address issues related to food safety, garbage collection, sanitation, vector control, and access to drinking water (Berg 2005). In the United States, the result was the most rapid reduction in morbidity and mortality (especially among children) in history (Berg 2005; Cutler and Miller 2005). With time, scientific knowledge evolved, containment measures became more sophisticated, and some infectious disease outbreaks were gradually brought under control with improved sanitation and the discovery of new vaccines (WHO 2007e).

Fourth Generation: The Invisible Regulator

By the middle of the 20th century, the institutional framework that regulates environmental health issues had become well established in the developed world (Berg 2005). Mandates and responsibilities for the different agencies involved in environmental health had been established. With dramatic improvements in environmental health conditions after World War II, the work of sanitarians evolved from more visible hands-on type of work to less visible regulation-based functions as health inspectors (Berg 2005).

Agenda Falling through the Cracks

The middle of the 20th century saw a rise in ecological awareness in the developed world, heralded by the publication and widespread reading of Rachel Carson's (1962) *Silent Spring* and the subsequent occurrence of numerous industrially related public health outbreaks. Growing concern about environmental exposures to industrial emissions led to the passage of numerous environmental laws and the creation of the independent environmental agencies. Increased emphasis on environmental laws and compliance ensued, and environmental health staff members were moved into these new environmental agencies. Over time, these staff members identified more with environmentalists, often forgetting their public health roots as they morphed into their new roles relating to environmental regulation (Kotchian 1997).

Their counterparts in state-level health agencies—epidemiologists and other health personnel—also began to lose their public health roles as budgets and agendas began to emphasize treatment over prevention. With a growing attention to health care delivery and hospital regulation, public health programs lost visibility and influence in the agenda of state health agencies (Kotchian 1997).

Decades later, the same pattern has been followed in developing countries. At one end is a growing environmental movement with the creation of ministries of environment and accompanying policies and regulations, while at the other end are health ministries and state health departments engaged in strengthening health systems and scaling up health sector programs that focus mainly on patient care. Environmental and infrastructure investments are not usually considered to be key determinants of human health by governments and health ministries— the latter tend to think of health investments in terms of number of hospital beds and doctors per person (Lvovsky 2001). In the process, the environmental health agenda has fallen between the cracks. It has become an institutional orphan as it is increasingly neglected in the programs of both the health and environment ministries.

Numerous other factors compound this relative neglect of the environmental health agenda in developing countries. High-level political commitment is a critical issue that is often lacking or, if it exists, does not get translated into effective policy discussions and actions. Additionally, both donor agencies and government investments work in isolated silos where investments in agriculture, nutrition, infrastructure, education, the environment, and health are often not well linked. Providing water and sanitation in urban areas, for example, has never enjoyed high priority for donors (Satterthwaite 2007), and funding for child health programs often includes only actions that lie within the health systems framework of the country. Country financing is usually allocated by sectors or ministries, so unless one sector takes the lead, no large-scale action can follow (Shekar and others 2006). Part of the problem stems from the fact that health ministries tend to focus on interventions in the health care system itself, rather than on primary prevention of environmental risks through infrastructural and behavioral interventions.

Institutional coordination at both the national and local levels is often weak. At the national level, poor intersectoral coordination across ministries means that environmental health issues—which are inherently cross-sectoral in nature—are not sufficiently addressed, and efforts remain ad hoc and often uncoordinated. Many of the environmental health interventions important for child health should be carried out at both the household and the community levels. In many developing countries, local government authorities—who are the best positioned to deliver environmental services and proper hygiene awareness—fall short in their efforts because of inadequate funding, weak revenue bases, and lack of staff members with expertise in environmental health. Local governments are also impoverished, powerless, and suffering from weak management capacity. In part, this state can be explained by the colonial legacy of imported codes and regulations for planning, land-use management, and buildings that were applied only to areas of the colonial city inhabited by "nonnatives." Also, however, the inadequacies of service provision can be attributed to a very rapid growth experienced around the time of independence (Satterthwaite 2007).

Finally, in many developing countries, the crumbling or inadequate infrastructure, underfinancing, and lack of personnel have eroded health systems even as the burden of disease has increased. Governments in developing countries find implementing and rehabilitating water infrastructure without donor support difficult. For example, in Central Asia, government budgets for operation and maintenance dropped from US$60 per hectare to less than US$8 per hectare. The result is that infrastructure is rapidly deteriorating (World Water Forum 2006). In the case of sanitation, more work needs to be done because 2.6 billion people lack adequate sanitation. Rapid urbanization and uncontrolled growth of urban slums are now creating a double environmental health burden for the urban poor. Poor urban people are exposed to health risks associated with dirty cooking fuels, traditional stoves, crowding, and poor access to water and sanitation (Satterthwaite 2007).

This neglect of environmental health within the national and local agendas of developing countries is critical—especially because the major environmental health concerns (poor water and sanitation access, inappropriate hygienic practices, lack of vector control, and continued indoor air pollution) are the same ones that developed countries faced about a century ago (Rosen 1958). Whatever the root causes for the poor environmental conditions and neglect of the environmental health arm of the primary prevention agenda by the health sector, the current reality in developing countries demands a revitalized approach to implementing environmental health interventions, given its importance for child survival.

Many governments in developing countries have increasingly begun to recognize the importance of environmental health in child survival, poverty reduction, and economic growth, and they are beginning to incorporate environmental health activities into existing integrated child development and nutrition programs. Additionally, successful community-level initiatives, such as the Total Sanitation Campaign in Bangladesh, are beginning to take root and to spread to other countries and regions (WSP 2005). In the following section, this chapter provides an illustrative list of different projects that have either recognized the need or incorporated environmental health into current policies, programs, and projects, such as child survival, nutrition, and water and sanitation.

Environmental Health Experiences in Developing Countries

Developing countries themselves vary considerably in terms of institutional capacity, political will, and socioeconomic development. Environmental health interventions therefore need to be customized to the specific enabling environment in a developing country. Recognizing these differences, rather than providing specific recommendations, this section presents examples of how some developing countries are beginning to mainstream environmental health components and objectives within existing child survival programs, nutrition initiatives, and infrastructure projects (water and sanitation, rural energy). As countries develop and their institutional capacity strengthens, they move toward more truly multisectoral programs that seek to address environmental health.

Integrated Child Survival Programs

In developing countries, several child survival strategies, such as immunization campaigns and improved case management of specific diseases, have proven to be effective. Nevertheless, accumulating evidence has suggested that child health programs need to move beyond addressing single diseases to addressing the overall health and well-being of the child. In response, during the mid-1990s, the World Health Organization (WHO), in collaboration with the United Nations Children's Fund (UNICEF) and many other agencies, institutions, and individuals, developed a strategy known as Integrated Management of Childhood Illness (IMCI).

Although the IMCI strategy mostly focuses on better treatment practices (in terms of diagnosis, medicine administration, and the like), elements of hygiene are already found within its key practices. Innovative projects in Cuzco, Peru, and Chinandega, Nicaragua, that have promoted key hygiene behaviors within the programmatic framework of their community-level IMCI strategies have resulted in significant improvements in health outcomes (Favin 2004). In another example, the Honduras Integrated Community-Based Child Care Program included a specific component on hygiene education, in addition to advice for improving home care, feeding practices, and related health behaviors (Griffiths and McGuire 2005). Similarly, the Integrated Child Development Services in India has included hygiene awareness (such as hand-washing practices for children) within a multi-sectoral approach that incorporates health, education, and nutrition interventions (Gragnolati and others 2006).

A few integrated child survival programs have been even more comprehensive—including components that traditionally lie outside the health sector but that have important health effects. For example, a subcomponent of the Eritrea Integrated Early Childhood Development Project included environmental health interventions to control childhood illnesses through promotion, maintenance, and use of safe latrines; improved food safety and hygiene; water source protection and handling; prevention of acute respiratory infections through indoor air pollution reduction interventions (for example, promotion of improved kitchen and household ventilation); and improved collaboration in the prevention of malaria and other vectorborne diseases (see World Bank 2000). Other child survival programs in low-income countries, such as Bangladesh, Benin, Brazil, Cambodia, Eritrea, Haiti, Malawi, Nepal, and Nicaragua, have included vector-control measures (use of insecticide-treated mosquito nets) as part of their preventive interventions (Victora and others 2005).

Some middle-income countries have begun to address child mortality more systematically by moving beyond health system interventions to a more integrated multisectoral program with strong political commitment from the government. Costa Rica, for example, added 15 to 20 years of life expectancy and significantly reduced child mortality in over two decades through a number of multisectoral policy actions, including a targeted program to control infectious diseases through improved sanitation, immunization, and the extension of primary health care to a wider population (World Resources Institute and others 1998). In Mexico, the government brought about significant improvements in under-five mortality by systematically rolling out programs on vaccinations, clean water, and health reform (see box 6.2).

Nutrition Programs

Improving the nutritional status of young children requires not only health sector interventions but also appropriate actions in agriculture, rural development, water supply and sanitation, social protection, education, gender programs, and community-driven development (Shekar and others 2006). Reviews of different nutrition programs in Central America, Ecuador, Thailand, Ghana, and India

BOX 6.2
Mexico: Multisectorality through a Diagonal Approach

Mexico is one of seven countries on track to achieve the goal of reducing child mortality by 2015 (Sepúlveda and others 2006). Improvement in children's health is attributed to two programs that worked together: the Clean Water Program and the Universal Vaccination Program. The Clean Water Program focused on appropriate water chlorination and regulations that, for example, banned the use of sewage for crop irrigation. Important steps in the program were to improve sewage-treatment plants, to provide adequate disposal of waste, and to monitor the maintenance of drainage systems.

The Universal Vaccination Program was created as a result of the measles epidemic in 1989 and 1990 (see the accompanying figure). The outbreak caused more than 70,000 cases and 6,000 deaths in mostly malnourished children under five years of age. Within three years of the project, immunization coverage exceeded 92 percent. By introducing national health weeks and improving the health monitoring system, the authorities focused on immunization, vitamin A supplementation, oral rehydration salts, and deworming—reaching sustained high population coverage. Evidence suggests that high levels of coverage of public health interventions as well as investments in women's education, social protection, water, and sanitation all contributed to the success (Frenk 2006; Sepúlveda and others 2006). This recent Mexican health reform included an unprecedented effort to strengthen environmental health services, regulatory actions to protect the public, and, more generally, a set of intersectoral interventions that define a health policy capable of modifying the broader determinants of disease (Frenk 2006).

Mortality in Children under Five Years, under One Year, and Neonates, Mexico, 1980–2005

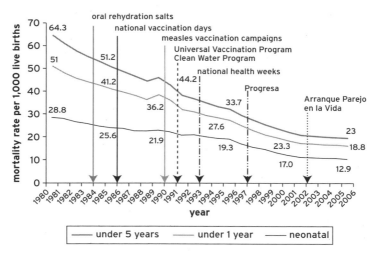

Sources: Data from the Mexican Ministry of Health Epidemiological and Statistical Mortality System and the Mexican National Council on Population.

illustrate the opportunities that exist to include or supplement environmental health activities and hygiene promotion in these programs.

Nutrition projects in developing countries also require complementary investments in water and sanitation; without these, children may lose the benefits of additional food intake to parasite infestation and diarrhea (Horton 1999). A review of health reforms in Central America found that programs addressing malnutrition should focus on preventive measures such as (a) ensuring access to safe water and sanitation; (b) educating mothers about nutrition, hygiene, and health practices; (c) using growth monitoring to ensure early detection of malnutrition; and (d) establishing links with health care specialists for at-risk populations, including the rural indigenous (World Bank 2007e). An analysis of India's main child survival program concluded that even with improvements in the Integrated Child Development Services, the country will not meet the Millennium Development Goals in nutrition without investing more in indirect interventions, such as water and sanitation infrastructure and women's education (Gragnolati and others 2006).

In the late 1980s, Thailand's nutrition program was able to effectively address malnutrition by adopting a more integrated approach (see box 6.3). In Ghana, the Community-Based Poverty Reduction Project (1999–2005, US$5.0 million) used an integrated community-based approach for basic health and nutrition service delivery. In addition to monitoring child growth, promoting pre- and

---

BOX 6.3
Thailand's National Nutrition Program

Thailand's National Nutrition Program successfully reduced moderate malnutrition from about 25 percent in the under-five population in 1986 to about 15 percent in 1995. Although the health sector led the nutrition program's outreach into villages, the fundamental assumption was that malnutrition was a symptom of poverty and, therefore, that a multisectoral attack on poverty was needed to deal with malnutrition. Accordingly, the Rural Development Plan and the Poverty Alleviation Program supplemented the nutrition program by funding a wide range of programs, including the following:

- Basic primary health and nutrition services
- Latrines and clean water
- A literacy program
- Agricultural production projects
- Village development projects, such as creation of fishponds or development of water sources
- Employment schemes during the dry season in disadvantaged, nonirrigated areas

Source: Heaver and Kachondam 2002.

postnatal care and family planning, and immunizing children, community workers advised on environmental sanitation and hygiene (World Bank 2007b).

An evaluation of programs in Ecuador (Sistema Integral de Alimentación y Nutrición) suggested that nutritional programs in rural areas be complemented with improved water and sanitation, community-based growth promotion, improved health service access, and nutritional supplements for pregnant mothers and young children, among other interventions. The study recommended that, at the national level, programs should include behavior change communications programs and work closely with the authorities responsible for water and sanitation (World Bank 2007f).

Infrastructure Programs

Development is often seen as synonymous with improvements in infrastructure, and accordingly, donor funding and government investments have been geared toward improving access to water supply, expanding coverage for sanitation and sewerage, and extending rural electrification. The result has been improvements in the access and coverage indicators for these environmental services. Despite these improvements in infrastructure, however, the trends in the incidence of diarrhea have not fallen correspondingly in developing countries (Kosek, Bern, and Guerrant 2003).

Over the years, donor agencies and country governments have increasingly realized that simply providing new infrastructure—be it latrines, piped water systems, or improved cookstoves—is not the "magic bullet" for achieving improved health. Elements of behavior change—the so-called software—are a critical part of the equation. For example, Peru's national government expressed interest in piloting a hand-washing campaign after US$2 billion of investments in water supply infrastructure failed to significantly reduce the prevalence of water-related diseases (Public-Private Partnership for Handwashing with Soap 2005). These behaviors—ranging from hand-washing practices to proper feces disposal to ventilation practices—are what make the difference between simply "having access" and actually "improving health."

One program that has brought together elements of infrastructure and behavior change has been the Total Sanitation Campaign in rural Bangladesh. This comprehensive program has been successful in eradicating the practice of open defecation by promoting behavior change and sanitation facilities in rural areas. Going beyond providing affordable sanitation options (in terms of technologies and costs), this program—as well as some similar programs—has been effective in ensuring high toilet use through a combination of participatory processes; hygiene promotion; institutional incentives, such as financial rewards for achieving universal toilet coverage; and community bans on practices such as open defecation (WSP 2005). The Total Sanitation Campaign has seen more than 400 villages achieve 100 percent sanitation coverage without a household subsidy for infrastructure (WSP 2005).

With the widely regaled success of the Bangladesh story, Total Sanitation Campaigns began to spread to Ethiopia, India, Nigeria, and Pakistan. The Ethiopian story was an equally successful one, with a strong multisectoral approach at the local level that emphasized education and community participation (see box 6.4). The effectiveness of the campaign varied across different states in India. Although these programs were meant to be based on the same principles of community involvement and attention to hygiene and social intermediation, a review showed that excessive attention to coverage targets overshadowed the goals of improved community health. Surveys of some villages showed that although they had 100 percent latrine coverage, several households were, in fact, using the latrines to store rice (WSP 2005).

With rural populations still largely dependent on biomass fuels for cooking and heating (and the consequent health impacts from indoor air pollution), country governments, donor agencies, and nongovernmental organizations (NGOs) have been working to help rural populations move up the energy ladder toward improved cookstoves and cleaner fuels (for example, liquefied petroleum gas). Although considerations of energy efficiency and costs have dominated rural energy technology choices, recent studies have pointed out the need to pay attention to changes in behavior. A research study in Bangladesh, for example, suggested changes in cooking arrangements and location, construction materials,

---

BOX 6.4
Ethiopia: The Toilet Revolution

Another example that drew from the Total Sanitation Campaign is in the Southern Nations Region of Ethiopia. The Regional Health Bureau, with support from the regional cabinet, shifted from a broad primary health care model to a focus on preventable diseases through a limited number of high-impact, broad-reach, low-cost public health interventions in rural areas. This change was facilitated by a strong regionwide commitment to empower households, resulting in a wave of household latrine building. At the village level, after consultation meetings highlighted the need to build latrines, community health promoters visited each household to persuade it to join the community in building latrines. Within less than two years, pit latrine ownership rose from 13 percent to 78 percent (WSP 2007), although it may not necessarily have resulted in a corresponding increase in use. The long-term success will depend on many factors, including durability, technical standards, design, and behavior (with regard to toilet use, and especially children's compliance with latrine use).

and ventilation practices that may help improve indoor air. The analysis also suggests that poor families may not have to wait for clean fuels or clean stoves to enjoy significantly cleaner air when basic modifications in cooking behaviors may produce cleaner conditions even when "dirty" biomass fuels are used (Dasgupta and others 2004a).

Vector-Control Programs

Malaria continues to be a major public health challenge, especially in Sub-Saharan Africa, where it is responsible for at least 700,000 deaths among children under five years of age (Snow and others 2003). Although key strategies in the control of malaria have often focused on early diagnosis and treatment, indoor residual spraying, and the use of insecticide-treated nets, environmental management interventions can play a major role within integrated vector management by adding resilience to the results of individual control strategies and by reducing costs as well as the likelihood of development and spread of drug and insecticide resistance (Keiser, Singer, and Utzinger 2005).

A review of 40 studies on malaria interventions that emphasized environmental management interventions reported very high protective efficacies. Often, a package of multiple interventions is used, guided by a multisectoral staff with expertise in malaria epidemiology and entomology, vector ecology, and land and water engineering (Keiser, Singer, and Utzinger 2005). In Malaysia, for example, the incidence of malaria was radically reduced by environmental management methods devised by interdisciplinary teams of medically qualified personnel across many different departments, such as agriculture, health, transportation, urban planning, and the army (Flemming and others 2004). Although environmental vector-control measures may work in Asia, they are much less likely to succeed in rural Africa. The African vectors are far more efficient, so their numbers must be reduced to much lower levels to reduce transmission significantly. For this reason, mosquito ecology demands that environmental manipulation be supplemented by bednets and indoor residual spraying (Touré 2001).

Lessons learned from successful vector-control programs show the need for interdisciplinary collaborations among the health, agricultural, water, and infrastructure development sectors in the design of integrated intervention packages. Collaboration, in this context, means that the higher expenses of environmental management are not borne by the health sector and that vector control per se is not the driving force (Keiser, Singer, and Utzinger 2005).

Effective vector-control programs should encourage community participation (Keiser, Singer, and Utzinger 2005). In the case of dengue control, a novel effort carried out in several communes in Vietnam demonstrates how a community-based approach can be highly effective against this mosquito (see box 6.5).

BOX 6.5
Vietnam's Dengue Program

The dengue-control case in Vietnam shows that a community-based approach can be highly effective against this disease. The approach involves a smooth interface among technical experts, health workers, and communities.

Dengue control was driven by the chair of each commune and by other leaders from the women's union and youth union; it was implemented by communal health personnel, health collaborators, schoolteachers, and their pupils. Health collaborators were each responsible for monthly inspection of water-storage containers in about 100 houses, delivery of health education messages to make sure that the householder knew how to manage the mesocyclops (a tiny one-eyed shrimplike creature used for biological control of larvae to combat mosquitoes successfully) that inhabited those containers, and reporting of any suspected dengue cases to the communal health center.

The program was not simply a matter of the health volunteers eliminating the mosquitoes for the householders—the success of the mosquito-control efforts depended on the close involvement of the house-holders. Schoolchildren participated by cleaning up discarded containers; supporting the aged and infirm in mosquito-control measures; and disseminating information about mosquito control through plays, songs, quiz nights, and—in one district—a Meso football cup.

This control strategy has already been implemented in 46 communes in Vietnam. Mosquitoes have been eliminated in 40 of those communes, and the remaining 6 communes show very low densities of mosquitoes.

Sources: Das Gupta and others 2006; Ha and Huan 1997; Kay and Nam 2005; Vu and others 2005.

Understanding the Enabling Environment

As the preceding section shows, developing countries have used a range of ways to mainstream environmental health actions into existing child survival and nutrition programs, infrastructure projects, and vector-control strategies. Given the varying socioeconomic, institutional, and developmental contexts of developing countries, knowing what combination of factors contributed to the successful inclusion of environmental health considerations within these programs is often difficult. Looking at the big picture, however, can help identify some common elements that provided the appropriate enabling environment. Some of these common elements have included obtaining high-level political commitment, involving and empowering communities, allocating responsibilities and resources at the local level, and finding a balance between private and public sector roles.

Political Will and Effective Consensus

The existence of a positive policy environment is a central principle for successful interventions. High-level political commitment is critical to successful child survival and environmental health programs. The Ethiopian example shows the need for a well-connected leadership that is well informed and willing to take risks and challenge conventional primary health care approaches. The right context is required to implement interventions. In the case of Ethiopia, the intervention was driven by population expansion and deforestation, which reduced the private open defecation option. The success of the story came from the willingness of senior management to listen to the people and put their needs first (WSP 2007).

An important first step in the program was facilitating consensus around the need for community ownership of health behavior—especially with respect to vaccination, family planning, and hygiene and sanitation. Consultation and advocacy meetings at all levels and covering the whole region were conducted to bring sustained massive response (Knapp 2006). As Shiferaw Teklemariam, state health minister in Ethiopia, points out, "[N]one of these achievements would have been possible without close inter-sectoral collaboration and strong leadership, committed to universal water supply and sanitation access" (WSP 2007: 2).

China's improved cookstoves programs have been found to be broadly successful in delivering better stoves to a majority of households in targeted counties, and the success was based on strong administrative, technical, and outreach competence and resources situated at the local level, motivated by sustained national-level attention (Sinton and others 2004). In Brazil, similarly, favoring a democratic and continuous process in education on health and disease processes has been found necessary in stimulating the participation of the population in health promotion programs (Mahendradhata and Moerman 2004).

Involving and Empowering Communities

Participatory governance mechanisms that encourage the involvement of poorer groups are an important factor for improving environmental health (Satterthwaite 2007). Several programs that have been successful in achieving child health outcomes through environmental health actions have included community participation as an integral component. In Ethiopia, community health promoters visited households, promoting simple and action-based health messages to undertake actions at individual, household, and community levels that would lead to behavioral change in practices (Knapp 2006).

Lessons learned from many programs highlight the need to cultivate a sense of local ownership. Community groups are potentially quite influential for behavior change and for maintaining infrastructure, as the Bangladesh sanitation story demonstrates (WSP 2005). In Thailand, effective consensus and commitment building in the multisectoral nutrition program was instrumental in reducing malnutrition among children (Heaver and Kachondam 2002). In Vietnam, the success of the dengue-control program also lay in the extensive community involvement (Das Gupta and others 2006).

Another example is the large program of toilet blocks in urban slums in India that were designed, built, and managed by the communities. The National Slum Dwellers Federation develops and promotes these blocks through a women's saving group (Mahila Milan) that is supported by local authorities, thus improving sanitation for millions of slum dwellers (Burra, Patel, and Kerr 2003). A key feature in Mumbai was the involvement of the slum community in project implementation at the planning stage. Their involvement facilitated collaboration between the NGO, contractors, and community-based organizations (WSP 2006).

Balancing Private and Public Sector Roles

The nature and complexity of environmental health issues require intersectoral actions not only across various levels of government but also through partnerships between the public and private sectors. While the public sector retains its power over setting regulations and standards, the private sector is increasingly getting involved in the delivery of environmental services such as water, sanitation, and energy. However, the question of who provides and pays for an intervention is important. Improved stoves are an example of private gains to public health measures. To be sure, the absence of credit is a possible constraint for individuals to obtain these stoves, but if the stoves save more on fuel than they cost, one would expect widespread adoption. In contrast, if the benefits are more general than individual, the scope for subsidies is clearer. An evaluation of costs and benefits is necessary to determine the role of the private and public sectors.

In the water sector, the privatization of water companies not only can lead to efficiency gains but also can have beneficial health effects. In Argentina, for example, the privatization of public water companies resulted in both increases in connections and improvements in quality (financed from user fees and connection charges). These improvements included the speed of repairs of water leaks and sewer blockages and the percentage of clients with appropriate water pressure. Child mortality fell by 8 percent in the areas that privatized their water services, and that effect was largest (26 percent) in the poorest areas, especially in poorer municipalities (Galiani, Gertler, and Schargrodsky 2005; Wagstaff and Claeson 2004).

In the field of sanitation, concepts of "sanitation marketing" are proving to be extremely effective. In a bid to harness market power for rural sanitation, IDE— an international NGO—developed a range of low-cost sanitation options and stimulated a network of local masons to market and deliver them to the rural population in two Vietnamese provinces. As a result, the sanitation access rate increased markedly in the area, even among the poor (Frias and Mukherjee 2005).

In hygiene behavior, the private sector has begun to play an important role in advocating for and supporting hand-washing campaigns through the Global Public-Private Partnership for Handwashing with Soap. In campaigns such as those in Ghana, Peru, and Senegal, the private sector has engaged in hand-washing promotion and educational activities mainly by contributing its marketing expertise to publicly funded generic campaigns not tied to any specific soap brands.

In providing rural energy technologies, too, the private sector is beginning to play an important role in collaboration with the public sector. In Nepal, for example, the Alternative Energy Promotion Center, a quasi-governmental body, is working closely with private sector contractors who deliver improved cookstove technologies to rural households (World Bank 2008b).

This section has provided examples of interventions in different sectors that have provided some key lessons; however, contexts, actors, and situations are different, which makes comparison difficult.

Governance and Institutional Implications

A final key issue is that environmental health efforts will not be sustainable unless explicit attention is paid to institutional issues. Therefore, the ultimate design and delivery of environmental health interventions must depend on the institutional and governance structure of each country, an issue that is elaborated on in this section.

Unraveling the Environmental Health Agenda in the Developed World

Earlier, this report discussed how, across the developed world, countries sought to establish strong and coherent systems for ensuring environmental sanitation to protect the public's health. As a result of those efforts, mechanisms to ensure environmental sanitation became well established, the burdens of associated disease dropped, and importantly, the public's expectations regarding the acceptability of poor sanitation and disease altered so that providing poor services has become politically very difficult. As mentioned earlier, recent decades have seen some significant unraveling of the coherence of the systems put in place over the previous decades. Environmental health has become disconnected from the mainstream agendas of the health agencies (which now focus more on a case management perspective) and the newly formed environmental protection agencies (which focus more on environmental regulation and compliance) (Kotchian 1997).

Correspondingly, donor initiatives related to environmental health issues have also become highly compartmentalized. With donor funding being channeled through separate sector silos (of environment, water and sanitation, and health), the possibility of generating the intersectoral coordination that is so crucial for environmental health initiatives was more or less preempted. This development is also evidenced in the way that international goals and development initiatives addressing health give very little attention to the role of environmental sanitation. Sanitation was excluded from the list of the original Millennium Development Goals and was only subsequently introduced (Satterthwaite 2007; von Schirnding 2005).

Developing Countries Follow Suit

As the world's poorer countries have moved along the development path with increasing investments in infrastructure, education, health, and other growth-oriented sectors, a similar trend has been seen, with government investments and donor funding increasingly programmed along sectoral lines. The role for the health sector is primarily that of regulation (for example, urban drinking-water quality); promotion (for example, hygiene promotion); and advocacy (for example, persuading utilities to lower their connection charges to the poor). In some cases, as in India (see box 6.6), attention by the health department to environmental health functions has atrophied over the years, as increased emphasis on treatment and delivery of health services led to a dwindling role for public health issues

---

BOX 6.6

Atrophy of Environmental Health Functions in India

The role of health departments in the stewardship of sanitation and public health activities has atrophied in India since the 1950s. Before independence, sanitary departments existed at national and provincial levels, with sanitary inspectors and other staff members on the ground in each district. In the districts, sanitary inspectors worked with the village officials and village *panchayats*, which had been set up to extend the bare bones of government to rural areas. The municipal areas also had extensive machinery for environmental sanitation and public health activities, including the management of water, solid waste, and liquid waste. Sanitary departments were accountable directly to the government and were administratively separate from the Indian Medical Service, which provided medical services.

In the 1950s, public health services were merged with the medical services into a unified health department. Interest in specializing in public health dwindled as specialty curative skills became far better rewarded. This atrophy was further fueled by strong electoral demand for services to cure the sick, while by contrast, public health services remained low profile because their success depended on ensuring no outbreaks.

Source: Das Gupta and others 2006.

(Das Gupta and others 2006). One cause for atrophy is the lack of demand for environmental health services. The public demands curative care over preventive care, and environmental health is the most invisible type of preventive care.

The consequences of such compartmentalization are much more severe than in the developed world, where intensive environmental sanitation measures had become fundamental to public administration by the middle of the 20th century. This subsequent compartmentalization may create inefficiencies and the reemergence of some threats, but the fundamental protections for public health remain firmly in place. By contrast, the era of compartmentalization arrived for many developing countries at a time when levels of exposure to environment-related infections were still very high and before a public health consciousness was intrinsically found in communities. Box 6.7 illustrates the evolution of environmental health institutions in Ethiopia, an example of such compartmentalization.

BOX 6.7
Institutional Evolution of Environmental Health: The Case of Ethiopia

Traditional Organization (1908–46)

The development of environmental health interventions in Ethiopia dates back to 1908, when it was organized under the Ministry of Interior. The Ministry of Interior consolidated decision-making power over health matters by proclaiming a series of legal notices following its mandate in sanitation. During this period, environmental health activities had an urban bias, the institutionalized aspect did not reach the grassroots level to address environmental health, and interventions were not significant because of limited human resources and unclear vision.

New Organization, New Tasks (1947–74)

In 1947, the Ministry of Public Health was established with a clear mandate, and rural service provision began to promote medical and sanitation services. The ministry set up a hygiene and environmental health unit for the control and prevention of communicable diseases, with a senior sanitarian guiding sanitation activities. A series of sanitation regulations was enacted, and a training center was set up to train sanitarians, who were largely involved in urban inspections pursuant to sanitation regulations. They were key actors in the eradication of smallpox in Ethiopia.

(continued)

Primary Health Care Emerges (Late 1970s–Late 1980s)

By the late 1970s, a socialist regime set up primary health care, which included drinking water and sanitation as formal components. A 10-year plan was made, with rural and urban programs used as entry points for sanitation efforts. Organized committees at various levels coordinated and ran environmental health interventions. The Ministry of Public Health was renamed the Ministry of Health, and in 1983, its Department of Hygiene and Environmental Health was organized into four divisions responsible for major sanitation activities: water and sanitation, food hygiene, industrial hygiene, and quarantine hygiene services.

Restructuring of Health Policy and Shift in Strategy (1990–Present)

The beginning of the 1990s brought a decentralization of powers down to the grassroots level and expansion of the private sector in environmental health services. Although the Department of Hygiene and Environmental Health maintained its status, regional states developed teams of environmental health professionals organized under the Department of Disease Prevention and Control or Department of Health Programs. Environmental health personnel worked in administrative units that coordinated and guided socioeconomic development at the grassroots levels. Health extension packages targeted households with a strong focus on community-based approaches and sustained preventive and promotional health care.

Ethiopia is a clear case of how an environmental health institutional framework has evolved over the years. In the most recent past, success has been largely due to the use of local government as a platform for environmental health, including hygiene promotion.

Source: Kumie and Ahmed 2005.

Compartmentalization at the national level means that the basic institutional mechanisms for successful governance of environmental sanitation are lacking at local levels. The consequences can be devastating. In Ghana, a government program for expanding irrigation through the construction of dams and canals, supported by the U.S. Agency for International Development, started as a well-designed effort to make water available for use by humans and livestock, as well as for agriculture and fishing (Hunter 1992, 2003). However, no measures were in place to protect against the spread of water-related diseases. Poor maintenance and absence of community extension services to encourage dredging, repairs, and vector control resulted in an increase in cases not only of schistosomiasis in children but also of filariasis among adults. The water supply project is described as a "classic example of un-coordinated unisectoral intervention by the Ministry of Agriculture and its international collaborators" (Hunter 2003: 231).

Institutional Requirements for Successful Environmental Health Governance

Successful environmental health governance requires strong institutional underpinnings, with clearly articulated roles at all levels of administration within a country. Because developing countries differ in their level of institutional complexity, this section talks in more general terms about the typical roles and responsibilities that should exist in countries for effective governance, but it caters to the more specific case of environmental health governance.[1]

Roles of National Governments

Successful environmental health governance is not possible without strong involvement at the national level.

Provide the regulatory framework. An appropriate legal and regulatory framework should underpin effective environmental health action. Many countries have such regulations, but many do not have adequate implementation mechanisms. In several African countries, for example, an examination of their legislation reveals an encouraging picture, with many countries having environmental health policies and sanitation codes (WHO Regional Office for Africa 2006), but problems with implementation remain. For example, in Ethiopia, issues with competing mandates (but also other institutional weaknesses) hinder implementation of reasonable policies. These regulations need to be backed by at least a rudimentary implementation framework, including mechanisms for building stakeholder compliance, monitoring mechanisms, and transparent enforcement mechanisms that minimize reliance on the overburdened judiciary.

These implementation frameworks, such as frameworks for ensuring the quality of public and private sources of drinking water, will serve to guide local governments to engage in appropriate environmental health actions. When necessary, local governments can amend model bylaws framed by the national governments, which can provide a regulatory basis for local environmental health activities, such as the maintenance and cleaning of drains and other civic amenities.

Provide the necessary platforms for intersectoral coordination. Environmental health is intrinsically multisectoral and requires intersectoral coordination across health, environment, infrastructure, and other sectors. In an era of sectoral ministries and silo mentality, intersectoral coordination mechanisms (such as intersectoral committees) need to be established at each level of the administrative hierarchy down to local governments. These mechanisms must be set up to work on a routine basis rather than in response to an emergency. They must involve all key departments whose actions pertain to environmental sanitation and public health outcomes. In the Philippines, for example, the Interagency Committee on Environmental Health was established in mid-1980s; the Department of Health serves as the chair, and the Department of Environment and Natural Resources is the vice chair. The committee has addressed priority environmental health problems in the country, such as water quality and sanitation, among other issues (Blössner and de Onis 2005).

Ensure interjurisdictional coordination. Environmental health exposures have significant public health externalities—so inadequate attention to clean water and proper sanitation, for example, results in infections such as diarrhea, which can cross jurisdictional boundaries. Interjurisdictional coordination therefore becomes essential, because local governments' efforts to reduce environmental exposures to infections can be seriously undermined if neighboring local authorities do not take similar action. Thus if a peri-urban area is poorly serviced, it can be a recurring source of disease outbreaks that can spread to the nearby metropolis.

Another set of issues arises because urban and rural local bodies have different regulatory, organizational, and implementation structures, and they typically have very low levels of coordination with each other. This situation greatly increases the probability of differential public health performance in neighboring localities and resultant externalities. Even within urban areas, ignoring the growing slum settlements with dismal environmental conditions and almost negligent access to environmental services can derail attempts by city governments to provide healthy environments.

Clarify mandates and appropriate devolution of authority and finances. Inappropriate devolution of authority is a perennial issue in any country, developed or developing, and becomes an issue in the implementation of environmental health interventions. For example, Dandoy (1997: 83) quotes a former U.S. state health officer as saying, "The federal government has most of the money, the state government has most of the legal authority, and the local government has most of the responsibility for protecting the health of the people."

Various institutional problems typically hamper effective environmental service delivery and reduce the scope for serious accountability. Problems relating to the decentralization of functions can result in delegation of environmental health functions to local governments without adequate administrative, financial, and technical support from line agencies or higher levels of government. In addition, insufficient clarity and differentiation of mandates and functions between the different layers of the government hierarchy and line agencies creates confusion, undermines accountability, and leads to duplication of efforts and waste of resources. Given the multisectoral nature of environmental health interventions, clear mapping of the specific roles of environmental, health, and infrastructural actors at each administrative level becomes even more critical. Finally, in many countries, funds may reside in higher levels of government than the one responsible for service delivery—creating shortfalls in financing.

Provide technical and policy inputs. Agencies at the national level play an important role in providing technical and policy inputs related to environmental health actions. Local governments typically lack the sectoral expertise on the wide range of issues they need to address to ensure environmental health, and they can benefit from technical advice and guidance from national-level agencies. One set of issues relates to providing the performance standards, norms, guidelines, training modules, and technical support for environmental health services. These require inputs primarily from the department of health, but also from other departments concerned with environmental management for health, such as the department of environment. In addition, national-level health agencies can provide critical information gathered from disease surveillance, assessment of health threats, and research to help local governments respond appropriately to environmental health risks.

Foster communication, advocacy, and promotion. Health ministries in developing countries have an important role to play in health promotion (Cairncross and Valdmanis 2006) to raise awareness among the public as well as in advocacy to raise the profile among policy-makers. Although nationwide campaigns on the need for immunizations (polio drives, for example) and public health messages relating to treatment of diseases (use of oral dehydration salts to treat diarrhea, for example) are commonplace, in only a few instances do national-level agencies play a role in promoting behavior change. Some examples of national hand-washing campaigns—such as Peru's "Lávate las manos con jabón" ("Wash your hands with soap")—have met with mass popularity and success in the adoption of better hygiene. Other examples include national hand-washing campaigns in Ghana, Nepal, and Senegal (for more information, see http://www.globalhandwashing.org). The health ministry can also play an advocacy role in raising the awareness of these issues within local and national governments. At ministerial meetings, it could link health outcomes with infrastructure projects, thus raising awareness of environmental health issues.

Roles of Local Governments

With guidance from national-level legal and regulatory frameworks, accompanied by devolution of functions and funding, local governments can play an important role in environmental health at the community and household levels.

Promote an integrated approach to local level health. Given the nature of environmental risks and transmission mechanisms, the most appropriate levels for carrying out actions are the household and community levels. Local governments are therefore best placed to ensure the intersectoral and interjurisdictional coordination needed to ensure public health, because they are the convergent point at the appropriate level of all the departments whose actions affect environmental health.

The benefits of intersectoral coordination at the local level are illustrated by a successful malaria-control program in Kheda district in Gujarat, India. Conducted with the active participation of local communities, it involved intensified case detection and treatment to reduce the number of mosquitoes carrying malaria parasites, coupled with eliminating vector breeding habitats in homes and in irrigation canals and the use of larvivorous fish. The active participation of the district and local governments and the high level of intersectoral coordination, including those responsible for the health, fisheries, and irrigation sectors, contributed to the success of this effort in reducing the number of mosquito breeding places (Malaviya and others 2006).

Provide civic amenities related to environmental health. A core task of any local government is to provide basic civic amenities, and local governments have at their disposal a range of funds and other resources that they can use for these tasks, including locally generated revenues. These basic amenities include the construction, maintenance, and cleaning of public resources, such as roads, drains, water supply systems, sewers, and waste disposal sites; measures to improve the sanitary condition of the locality, such as solid waste removal; and monitoring activities, such as the protection of water sources.

Mobilize citizens and support health activities. With their strong grassroots presence, local governments can do a great deal to enhance the effectiveness of services and programs aimed at improving health. For example, they can help collect disease surveillance information by reporting easily identifiable conditions, such as an episode of severe vomiting and diarrhea. They can disseminate public health messages and mobilize citizens to improve their own health behaviors and environmental hygiene and to pressure others around them to help maintain better sanitary conditions in their neighborhoods.

Additionally, local governments can mobilize community-based groups to help improve environmental services. The total sanitation program in West Bengal illustrates how successful local governments can be in inducing widespread adoption of personal latrines and in using NGOs to set up for-profit "sanitary marts," where latrines are produced to given standards and retailed for an agreed price (Majumder 2004; also see discussion in Mavalankar and Manjunath 2004). Under the leadership of local governments, citizens' groups such as self-help groups are active in the rural areas of West Bengal and have proved to be a powerful force in the spread of pit latrines. These successful efforts to mobilize citizens can be extended to a larger range of environmental health issues, such as solid waste management, vector control, and drainage.

Strengthen roles in communication and advocacy. A critical component of environmental health programs is to raise citizen awareness of the benefits of clean water, proper sanitation, and good hygiene behavior, as well as to stimulate their demand for these services. Although national-level campaigns play an important role in raising awareness, more community-specific and customized messages relating to behavior change for better environmental health should be advocated by local health units and schools. Local governments can reach citizens more effectively by engaging communities and individuals in a host of face-to-face forms of communication.

Reorient the role of an environmental health workforce. Whether under the aegis of the health department or the local governments, a minimum complement of public health personnel is needed at each administrative level to implement and ensure good environmental health. Unlike a century ago, the all-encompassing role of the environmental health officer may no longer be realistic. The role will have to be reoriented through in-service training at the local level to address the changing priorities in environmental health. In South Africa, for example, environmental health officers form the backbone of environmental health services and are employed at the local and state levels. Because of the need for multiple skills in response to the crosscutting nature of environmental health interventions, training of environmental health officers has changed fundamentally, with the focus moving beyond law enforcement to community participation and development (Mathee, Swanepoel, and Swart 1999; Thomas, Seager, and Mathee 2002).

Establish and enforce regulations. Local government can play an important role in establishing, monitoring, and enforcing regulation. This activity involves building codes, land tenure bylaws, and tenancy regulation. Regulations can require that no house be occupied without prior construction of a latrine, can prevent landlords from raising rent if the tenant has a latrine built, and can even show the mason how to build it. Local government is usually responsible for outbreak investigation and the ensuing environmental health measures, with the cooperation and technical knowledge of experts in the national governments. Environmental health regulations are necessary but not sufficient, because many environmental health issues are the result of informal settlements or slums, where regulations are not enforced.

The roles and responsibilities at the local and national levels discussed here are generic and broadly represent the key institutional requirements for better environmental health governance. An important next step would be to carry out an in-depth institutional analysis at country level to better understand the coordination mechanisms between ministries and how mandates and budgets are assigned. Such an understanding is critical for environmental health, an issue that is inherently multisectoral, requiring collaborations across health, environment, and infrastructure ministries. For a specific country, this type of analysis would help governments better define the roles and responsibilities of different agencies, correct the incentive structures, and improve the implementation of environmental health programs and actions.

A Critical Moment

A critical moment has arrived for this neglected story to take the forefront. As this chapter shows, governments in several developing countries are beginning to incorporate environmental health actions in their sectoral strategies. Similar changes are occurring in the donor community, with a growing recognition of the crucial role of environmental health. Multilateral development banks such as the World Bank—acknowledging the need for multisectoral collaborations to improve health outcomes—are revisiting the role of the health sector, while acknowledging the multisectoral nature of such outcomes. The Bank's newly released Health Sector Strategy states:

> One of the clearest cases for strong government intervention in the [health, nutrition, and population] sector can be made when there are large externalities (the benefits to society are greater than the sum of benefits to individuals). This is true in the case of clean water, sanitation services, vector control, food safety measures. . . . (World Bank 2007c: 5)

Additionally, other international organizations such as WHO are strengthening measures to evaluate the disease burden from environmental risks in developing countries. Recognizing the effect of environmental health on malnutrition, new WHO burden of disease estimates are now including the consequences of malnutrition that are due to poor sanitation in developing countries (Fewtrell and others 2007). The Pan-American Health Organization stresses that a multisectoral community health approach is vital to reduce many of the "neglected" diseases in Latin America, which are often preventable through improved environmental health (Holveck and others 2007). UNICEF, with its focus on child survival and development, has been consistent in its approach emphasizing water and sanitation measures alongside nutritional and health system strategies.

Key Messages

- Lessons from history have shown the enormous benefits of multisectoral environmental health actions. Today's developed countries have undergone an evolution in environmental health functions. Both institutionally and conceptually, however, environmental health has fallen through the cracks in the development agenda in the world's poorest countries.

■ Several governments in developing countries are beginning to mainstream environmental health components and objectives within existing child survival programs, nutrition initiatives, and infrastructure projects (water and sanitation, rural energy).

■ Some common elements for successful environmental health actions in developing countries have included obtaining high-level political commitment, involving and empowering communities, allocating responsibilities and resources at the local level, and finding a balance between private and public sector roles. Furthermore, successful environmental health governance requires strong institutional underpinnings, with clearly articulated roles at all levels of administration within a country.

■ Now is a critical moment for this agenda to take the forefront in developing countries, with governments, donors, and civil society beginning to strengthen measures to address environmental health, especially in the context of child survival.

Note

1 The framework provided in this section is based on Das Gupta and others' (2006) analysis of environmental sanitation and public health in India and was adapted to discuss the broader issue of environmental health.

CHAPTER 7

Conclusion

THE CHILD HEALTH AGENDA REMAINS UNFINISHED in the developing world, with millions of children continuing to fall sick and die from preventable environmental health causes. Malnutrition, poor environmental conditions, and infectious diseases are highly associated geographically and take their heaviest tolls on children under five years of age in South Asia, Sub-Saharan Africa, and certain countries in the Eastern Mediterranean region (Fewtrell and others 2007).

The contribution of repeated infections, especially diarrheal episodes, to a child's nutritional status and subsequent mortality has only recently been incorporated in assessments of the environmental burden of disease (Fewtrell and others 2007). A recent collective expert opinion stated that about 50 percent of the consequences of malnutrition are in fact caused by inadequate water and sanitation provisions and poor hygienic practices (Prüss-Üstün and Corvalán 2006). Evidence from a review of recent cohort studies undertaken for this report further corroborates this link between environmental infections and malnutrition.

Contributions of This Report

This report reemphasizes the window of opportunity highlighted in the World Bank's 2006 nutrition report—pointing to a critical period in the life cycle of a child, from the womb to the age of about two years (World Bank 2006c). With research evidence from various studies, this report demonstrates how exposures to environmental health risks jeopardize the health and nutritional status of young children. Poor environmental management and bad sanitation expose pregnant women to both malaria and hookworm infections. Left untreated, these infections lead to permanent growth faltering, lowered immunity, and increased mortality for these women's children. In early infancy, improper feeding practices and poor sanitation have a pernicious synergistic effect on the child's nutritional status. Many of these impacts on a child's growth also result in negative cognition and learning impacts as well as chronic diseases later in life.

Although considerable progress has been made, this report calls attention to the potential of untapped environmental health actions to complement existing health, infrastructure, and environmental management strategies in the developing world. Current child survival strategies in developing countries focus on case management and treatment, neglecting primary prevention, especially as it relates to reducing exposure to infections. The evidence of the impacts of environmental risks and subsequent infections on child malnutrition makes correcting the current neglect of environmental health in child survival and child health strategies even more imperative in developing countries.

Environmental health actions supplement existing strategies addressing child health. An analysis of results from Fewtrell and others (2007) shows that environmental health interventions have a multiplier effect on mortality: investments addressing environmental risks (for example, lack of water and sanitation) pay off with more than just their direct effect on disease (for example, reducing diarrhea). This finding has important consequences for developing countries seeking to invest in environmental health interventions; their budgets will stretch further because of the nature of the externalities associated with such interventions. Therefore, governments in developing countries have an enormous opportunity to incorporate environmental health interventions in existing strategies that affect child health and to provide a bigger "bang for the buck" in terms of health improvements. As has been discussed in this report, environmental health actions add value to existing programs for child care, micronutrient supplementation, and immunization. They can also be used to adapt environmental management programs in developing countries, such as those relating to vector control, as well as to form the basis of adjustments to strategies in the water sanitation and rural energy sectors to enhance health outcomes.

Revised country-level estimates show that when malnutrition-mediated health effects attributed to environmental health risks are included, the total costs are at least 40 percent higher for both Ghana and Pakistan. The longer-term effects of malnutrition (partly attributed to environment-related infections) on education and income are estimated to add another 3 to 4 percent in terms of annual costs. The total annual cost attributed to environmental health risks, including all malnutrition effects (such as higher education costs), are estimated to be as high as 9 percent of Ghana's or Pakistan's gross domestic product. Given this considerable social and economic burden, policy-makers in developing countries where malnutrition and poor environmental conditions coexist should place environmental health interventions high on the policy agenda. Furthermore, as developing countries strive to meet several of their commitments for the Millennium Development Goals, these interventions will remain critical for addressing poverty reduction and child mortality.

Growing evidence indicates the need to mainstream environmental health interventions in developing countries' child survival strategies. With environmental health having fallen through the cracks in the development agenda, however, policy-makers in those countries will have to make a renewed effort to revitalize the environmental health content within existing child survival programs, nutrition initiatives, and infrastructure projects. Anecdotal evidence of such efforts is beginning to emerge in developing countries. Common criteria for successful implementation include obtaining high-level political commitment, involving and empowering communities, allocating responsibilities and resources at the local level, and finding a balance between private and public sector roles. Finally, successful environmental health governance requires strong institutional underpinnings. This report discusses—through an institutional lens—the roles that national and local governments can play in delivering and managing environmental health interventions.

Next Steps

In many ways, this report represents a first step toward providing policy-makers with the epidemiological, economic, and experiential evidence to incorporate environmental health in the child survival agenda. However, in each of the three main sections of the report, additional research and studies will help donors and governments in developing countries choose to invest in appropriate environmental health interventions.

The epidemiological evidence presented in this report reaffirms the importance of environmental health for child survival, especially when considering the links through malnutrition. Further research on the environmental health impacts during pregnancy, additional disease transmission pathways, and better relative risk estimates will help improve disease burden and costing estimates while informing governments about appropriate programs. For example, in Sub-Saharan Africa, malaria and hookworm infections coexist to cause anemia and subsequent growth retardation of the fetus in pregnant women. Communitywide investments to improve water resources management, use of insecticide-treated nets and indoor residual spraying, and improved sanitation facilities would help improve birth weight.

This report demonstrates the substantial burden and consequent economic costs associated with environmental health risks, but important questions remain: How should governments prioritize among the different environmental health, infrastructure, nutrition, and child survival interventions to improve child health? What are the cost-benefit ratios and the levels of cost-effectiveness of individual interventions? Some research on this subject has provided answers at the global level—with hygiene promotion measures emerging as the most cost-effective intervention (Laxminarayan, Chow, and Shahid-Salles 2006). Carrying out such cost-effectiveness and cost-benefit analyses at a country level is the next step in guiding decision-makers in specific countries in prioritizing among the various available interventions. Some recent analyses in Colombia and Peru have begun to explore the ranking environmental health interventions through cost-benefit analyses (World Bank 2006a, 2007g).

In addition to conducting economic analyses, governments in developing countries need to look toward implementing specific environmental health actions. The effectiveness of such interventions ultimately depends on the enabling environment in the country, which highlights the need to assess (a) institutional mandates and capacities for addressing environmental health issues, (b) regulations on environmental health, and (c) availability of budgets (Poverty Environment Partnership 2008). However, a one-size-fits-all solution will not work, and policy-makers will need to look beyond the generic roles and responsibilities for better environmental health governance discussed in this report. An important next step would be to carry out an in-depth institutional analysis at a country level to better understand the coordination mechanisms between ministries and the ways mandates and budgets are assigned. Such an understanding is critical for environmental health—an issue that is inherently multisectoral and that requires collaborations across health, environment, and infrastructure ministries.

Temporality is also an important consideration in how governments choose between various interventions that address child survival, because the costs of programs are often incurred in the short term and are front-loaded, whereas the benefits—especially those relating to cognition and learning—are spread over the long term. In the absence of appropriate economic analysis at the country level, governments may be tempted to choose interventions that are cheaper in the short term, such as oral rehydration programs, rather than the more expensive water and sanitation programs, even though the latter may have greater benefits over the long term.

Additionally, over the longer term, environmental health concerns are expected to grow. As the world's climate changes, diseases such as diarrhea and malaria, among other important health burdens that are the result of environmental risk factors, are likely to worsen, particularly for the poor and in developing countries (Campbell-Lendrum, Corvalán, and Neira 2007; IPCC 2007). Changing temperature and precipitation will also affect agricultural production and threaten food security, thus having implications for malnutrition. Therefore, scaling up preventive environmental health interventions (such as clean water and sanitation) to reduce the current burden of disease is a prudent investment (Campbell-Lendrum, Corvalán, and Neira 2007).

Given the multisectoral nature of environmental health issues, the advocacy and regulatory role of the health sector and the supporting roles of other sectors (for example, environment, infrastructure, agriculture, and education) in promoting and delivering environmental health actions need to be revitalized. Without focused and targeted attention to improving environmental health, governments and donors are losing out on the significant opportunities to make dramatic changes in the child health scenario in the world's poorest countries.

Ultimately, good environmental health governance is about how policy-makers in the poorest countries develop mechanisms to pick up signals on environmental risks and then find ways to translate these findings into appropriate and well-targeted actions. These governments will also need to adjust their policies to address environmental health outcomes and set up institutional mechanisms to successfully implement interventions. Finally, the creation of long-term constituencies within a country will help to continually raise attention to issues related to the environment, health, and poverty and to promote social accountability among public officials for effective action on these issues (Poverty Environment Partnership 2008).

A concerted and continuous effort is needed on behalf of both developed and developing countries to ensure that environmental health is placed high on the development agenda and that corresponding interventions are financed and undertaken to improve children's survival and development potential.

APPENDIX A

Technical Review of Cohort Studies

Background

The World Health Organization (WHO) monograph by Scrimshaw, Taylor, and Gordon (1968) initiated a lengthy discussion on whether repeated infections in childhood resulted in permanent growth faltering and stunting in adulthood. Results from observational cohort follow-up studies on effects of infections over time—and ultimately over generations—have provided evidence to stimulate the debate.

By the time the WHO monograph was published, extensive microbiological, immunological, and physiological studies—partly based on human experiments—favored the idea that all infections contribute to growth faltering, but the long-term effects of infections on nutritional status were still lacking (Scrimshaw 2003). Diarrheal diseases are very common from infancy up to early childhood. Unlike upper respiratory infections, which are relatively uncommon in early infancy because of maternal antibodies (Nair and others 2007; Sato and others 1979), diarrheal diseases affect the gastrointestinal tract of the child from birth onward, causing malabsorption. For this reason, diarrheal diseases have been particularly studied to investigate their impact on growth faltering (Scrimshaw 2003; Stephensen 1999).

Search Strategy and Selection Criteria

For this study, an initial review of cohort studies published on Pubmed that linked infections and malnutrition (underweight, wasting, and stunting) was undertaken. In all, 38 cohort studies conducted in developing countries (table A.1) were identified; one recent and very large case-control study in Bangladesh (Chisti and others 2007) was also included. Studies that reported the effects of infections on underweight status (weight for age) only—that is, those that lacked information on stunting (measured by height for age)—were also included. One cohort study that was included described only the natural growth history of patients with persistent diarrhea (Valentiner-Branth and others 2001) and had no control group. One study merely descriptively associated diarrhea episodes and development of underweight (Schorling and others 1990). These studies were included whether or not associations of diarrheal and other diseases with malnutrition were separated.

Published experimental studies (and their systematic reviews) that report results from deworming trials undertaken after a recent meta-analysis (Dickson and others 2000) were separately included to allow inferences to be drawn about whether the elimination of soil-transmitted helminths that cause loss of appetite, malabsorption, or nutrient losses affect the nutritional status of individuals (Taylor-Robinson, Jones, and Garner 2007).

Findings and Discussion

Epidemiological studies typically examine associations between an exposure variable and a health outcome. Austin Bradford Hill recognized the importance of moving from association to causation as a necessary step for taking preventive action against environmental causes of disease (Lucas and McMichael 2005). Hill set out nine viewpoints, or guidelines, against which an epidemiological association might be assessed when attempting to reach an appropriate conclusion on causation: strength of association, consistency of results, specificity (dose-response), temporality, biological gradient, plausibility, coherence, experiment, and analogy. In this review of the literature, some of the more important of these criteria are used as a framework for evaluating whether infections cause growth faltering.

Strength of Association

The cohort studies use heterogeneous analyses and different parameters to study infection-attributable growth faltering. One often finds somewhat weak associations in these studies. One should bear in mind the following issues, which lead to difficulties in establishing an infection-attributable fraction of poor growth:
1. The internal comparison group—or merely a diarrhea-free period—in the reviewed studies is often not infection free. Thus, asymptomatic children may

have endemic diarrheagenic bacteria (12 to 37 percent, according to Nguyen and others 2005). Schorling and others 1990; Steiner and others 1998), as well as carry parasites and protozoa (Checkley and others 1997; Prado and others 2005). These children have been shown to present with pathological damage in the gut, which results in subsequent growth faltering (Checkley and others 1997; Steiner and others 1998). A cohort follow-up study of 71 Gambian infants has provided evidence that such damage—which results from a malabsorption disease likely caused by unhygienic conditions (Lim 2001)—causes 55 percent of the loss in weight gain in developing countries, even if infants were breastfed (Campbell, Elia, and Lunn 2003). In an earlier study in rural Gambia, infants between the ages of 3 and 15 months were found to have this malabsorption in the gut for 75 percent of the time but showed diarrheal symptoms only 7.3 percent of the time (Lunn, Northrop-Clewes, and Downes 1991).

2. In earlier studies in Guatemala and Nigeria (Martorell, Yarbrough, and others 1975; Morley, Bicknell, and Woodland 1968) and in a more recent study in Malawi (Maleta and others 2003), the comparison group was also exposed to a relatively high morbidity burden, which obviously diminishes the association between growth faltering and infections.

3. In most studies, the children being studied are subject to nutritional interventions (especially breastfeeding promotion) or receive appropriate medical case management. These interventions and treatments are also likely to mask the effects of diarrhea on growth because it becomes difficult to observe the natural history of growth faltering attributable to infections under these circumstances.

4. Another reason for difficulties in attributing growth faltering to infections lies in the methodology for research studies. Many studies (for example, Briend and others 1989) do not start follow-up from early infancy. This lessens the ability of investigators to find the effects of infections on growth faltering (Checkley and others 2003; see also the discussion of early infancy in chapter 2).

5. Furthermore, the environmental effects in the fetal period are not adequately considered in these cohort studies. Many infections can cause intrauterine growth restriction (Graham and others 2006), but recent robust ecological (van Geertruyden and others 2004) and analytic (Watson-Jones and others 2007) evidence reveals that malaria during pregnancy has dramatic adverse effects on fetal and newborn survival and related nutritional status at birth, especially in Sub-Saharan Africa.

Consistency

Infections contribute to growth faltering. This proposition is supported by 35 of the reviewed cohort studies. In the three studies in which the effects were largely not found (Alvarado and others 2005; Cohen and others 1995; Kolsteren, Kusin, and Kardjati 1997), the results can be explained by the well-established masking effects of breastfeeding (Brown 2003).[1]

Studies by Briend and others (1989) and later by Moy and others (1994) paid attention to the timing of infections within the follow-up periods. (Growth periods were analyzed as statistical units rather than by individual children.) The results demonstrated that the effect on growth faltering was observed to be corrected by catch-up growth among children experiencing watery nonchronic diarrhea if enough time had elapsed since diarrhea episodes. However, even those studies have agreed that chronic diarrhea (that is, diarrhea that persists for greater than 10 percent of time) has irreversible effects. The study by Briend and others (1989) also suggested that dysentery adversely affected linear growth after a three-month lag.

Studies that started follow-up from birth (see also the specificity and temporality arguments later) have shown with increasing consistency that diarrheal episodes lead to irreversible growth faltering. The Gambia study (Campbell, Elia, and Lunn 2003: 1337) concluded that "the lack of association between diarrheal disease and long-term growth as described by Briend et al. (1989) does not preclude the possibility of the presence of a growth retarding gastrointestinal enteropathy." Hence, when the gut of young infants is damaged from malabsorption disease, growth faltering will result even in the absence of diarrheal symptoms (and may not be attributed to the diarrheal episodes). Investigators have already questioned how children in the Bangladesh study (Briend and others 1989) could have caught up in growth despite the extra nutritional requirements generated by this common malabsorption disease (Lunn, Northrop-Clewes, and Downes 1991).

There are few follow-up data concerning helminths. One study from a northeastern Brazilian shantytown, where subjects were followed from birth to seven years of age, showed that helminths, independently and additively, caused 4.6 centimeters of growth faltering, on average. This result was on top of the effects of diarrhea (Moore and others 2001) and is consistent with an early quasi-experimental study from rural St. Lucia, West Indies. That study revealed that the introduction of household connections and latrines resulted in lower levels of *Ascaris* and *Trichuris*, lower diarrhea prevalence rates, and improved growth in infanthood (Henry 1981).

Specificity (Dose-Response)

Researchers have shown dose-response relationships between infections and growth faltering in several ways that closely relate to the specificity criteria:
1. Four studies (Adair and others 1993; Brush, Harrison, and Waterlow 1997; Checkley and others 2003; Moore and others 2001) have demonstrated that diarrheal

episodes (the dose), with increasing period prevalence, show lagged or lasting effects in terms of growth faltering (the response). This finding is consistent with many older studies that performed regression analyses for effects that did not lag (Baumgartner and Pollitt 1983; Black, Brown, and Becker 1984; Lutter and others 1989; Martorell, Yarbrough, and others 1975; Rowland, Cole, and Whitehead 1977; Zumrawi, Dimond, and Waterlow 1987).

2. Several studies have demonstrated dose-response effects relating the nature of infection bouts to growth faltering: the more severe the infection bouts, the more significant the growth-faltering effects in children. Especially in the case of invasive infections (such as enterotoxigenic *E. coli, Shigella,* measles, and acute lower respiratory infection), significant metabolic (or merely dose-response) effects have been observed (Alam and others 2000; Black, Brown, and Becker 1984; Briend and others 1989; Chisti and others 2007; Henry and others 1987; Victora and others 1990; Villamor and others 2004), as expected by Powanda and Beisel (2003) and Scrimshaw, Taylor, and Gordon (1968). Even the study that has most strongly challenged the irreversible effects of infections on growth makes an exemption with respect to dysentery (Briend and others 1989).[2] Also, infections (for example, protozoa) pose special immunological challenges in early, preweaning infancy (Marodi 2006; Wilson 1986) appear to contribute to permanent growth faltering (Checkley and others 1997, 1998; Molbak and others 1997).

3. Prado and others (2005) show that among *Giardia*-infected children, the more time the child remained untreated during the follow-up period, the more he or she experienced lagged linear growth faltering.

Temporality

The proposition that the associations between infections and growth faltering are merely transitory has been given importance in several relatively new studies, which emphasize that there are lagged or lasting effects that continue until two to seven years of age if the burden of infection began in early infancy (Adair and others 1993; Checkley and others 2003; Maleta and others 2003; Moore and others 2001; Prado and others 2005). This interpretation is supported by many earlier as well as more contemporary studies (Baumgartner and Pollitt 1983; Brush, Harrison, and Waterlow 1997; Checkley and others 1997; Chisti and others 2007; Fikree, Rahbar, and Berendes 2000; Martorell, Habicht, and others 1975; Martorell, Yarbrough, and others 1975; Molbak and others 1997; Morley, Bicknell, and Woodland 1968; Rowland, Cole, and Whitehead 1977; Victora and others 1990).

As follow-up data from Central America indicate, the growth status of children at two years of age strongly predicts their height in adulthood (Martorell 1995). The observed lagged or lasting effects are considered irreversible (Checkley and others 2003). These lagged effects are also important because they reduce the possibility of reverse causality (that is, that malnutrition increased the risk of infection in the first place) (Prado and others 2005).

Biological Plausibility

The proposition that infections cause undernutrition is biologically plausible. There are many mechanisms, both proven (anorexia, malabsorption, loss of nutrients) and still hypothetical (decreased nutrient transport to periphery), that interact with the host's immune responses (table 2.2; see also Keusch 2003; Scrimshaw 2003; Scrimshaw, Taylor, and Gordon 1959, 1968).

An important but surprisingly little-studied mechanism is the acute phase response to infection in children. The withholding of food (because of anorexia) in the acute phase of an infection such as diarrhea appears initially to have some beneficial effects. However, long-lasting anorexia delays recovery and is ultimately deleterious to a child's health (Langhans 2000). This mechanism probably underlies widespread traditional cultural practices relating to the deliberate restriction of food intake in children during disease episodes. It was only in 1997 that continued feeding during diarrhea was included as an important component of diarrheal patient management in the American Academy of Pediatrics' *Pediatric Nutrition Handbook* (Brown 2003).

Experimental Evidence from Deworming

Deworming with anthelminthics is often safe and highly effective in reducing or eliminating such soil-transmitted helminths as roundworm (*Ascaris lumbricoides*), hookworm (*Necator americanus* or *Ancylostoma duodenale*), or whipworm (*Trichuris trichiuria*) (Awasthi and Bundy 2007; Dickson and others 2000). WHO recommends[3] the widespread use of anthelminthics among children, in an effort to cut transmission that occurs in tropical countries (Awasthi and Bundy 2007). More and more, in large areas of developing countries, almost all children—especially after weaning—are being infected (Awasthi, Bundy, and Savioli 2003).

Trials carried out in the late 20th century that tried to establish the effectiveness of anthelminthics in improving nutritional status were conducted largely on schoolchildren infected with roundworm. The trials showed that, despite methodological problems, a single anthelminthic dose may lead to a significant weight gain of 0.24 kilograms (95 percent confidence interval, 0.15 to 0.32 kilograms; fixed-effect model) to 0.38 kilograms (95 percent confidence interval, 0.01 to 0.70 kilograms; random-effect model) over a period of one year or more (Dickson and others 2000). Subsequent rapid reinfection often masks the beneficial effects of deworming on weight gain.

A recent comprehensive meta-analysis of randomized trials supports the idea that a single dose of a deworming drug has a significant effect on the mean change in weight in children who have an initial high worm prevalence or intensity, compared with no effect in a group with low worm prevalence or intensity (Taylor-Robinson, Jones, and Garner 2007).

Conclusions

Consistent and biologically plausible results from the reviewed cohort studies, coupled with evidence of specificity, reveal how increments in infection load—number, severity, and duration of episodes—are associated with stronger effects on nutrition, which appear after a time lag in terms of stunting. The results support the idea that infections—especially if acquired in early infancy—cause irreversible and often highly prevalent undernutrition among children under five years of age in developing countries.[4] Experimental anthelminthic trials on humans also support the idea that soil-transmitted helminths and other pathogens associated with similar pathophysiological effects (malabsorption and loss of nutrients) cause undernutrition.

In summary, several strong masking elements are important in estimating how much of undernutrition is attributable to infections:

1. Enteric infections cause growth faltering in children in developing countries, even in the absence of diarrhea. These infections make children with diarrhea and without diarrhea appear to be more alike with respect to growth.
2. In some studies, there is a relatively high disease burden in the control group, which again makes children with and without infections more alike.
3. The active case management and supporting nutritional therapies (including breastfeeding) that are often targeted at study participants mitigate the effects of infections on nutritional status.
4. Not enough infants are enrolled in many of these cohort studies; therefore, the studies miss the long-lasting effects in the most vulnerable period (0 to 6 months) on nutritional status.
5. Environmentally attributable growth faltering (especially malaria and hookworms) during the fetal period that manifests itself in intrauterine growth retardation is not adequately captured in these studies.

Independent of this review, a group of experts chosen by WHO (Prüss-Üstün and Corvalán 2006) estimated on the basis of converging expert opinion (using the Delphi method) that 50 percent (with a confidence interval of 39 to 61 percent) of the consequences of maternal and childhood underweight are attributable to environmental issues, especially to poor water, sanitation, and hygiene. Fewtrell and others (2007) use this attributable fraction in estimating the regional burden of disease from poor water resources management (including lack of water, sanitation, and hygiene). The regional estimates of disease burden presented in chapter 4 are based on those two studies.

For reasons such as heterogeneous inclusion criteria and analysis techniques and differing follow-up times, masking effects, and the like, it is virtually impossible to assess the literature systematically to find out how much (what percentage) of growth faltering is attributable to diarrheal episodes. Furthermore, not all studies present their results; hence, the variance of growth faltering that is explainable by infections is not readily available.

It is important to note that because the prevention of infections more than proportionately moves children from a severe to a moderate and mild malnutrition state (see table D.5), far less than 50 percent of weight gain attributed to environmental infections explains 50 percent of the consequences of maternal and childhood underweight. The idea that a substantial proportion of children with severe undernutrition can be lifted to the moderate and mild categories is also supported by experimental data on humans (Ahmed and others 1993).

Findings from this review of 38 cohort studies support the 50 percent figure used by WHO in its reports. Examples of cohort studies from this review provide some idea of how experts have come up with this figure. For example, a recent relatively large study in Bangladesh reveals that dysentery and watery diarrhea together could retard weight gain by 20 to 25 percent compared with the results in periods when no infections occurred (Alam and others 2000). The weight gain retardation is likely to be significantly higher when compared with international standards. Simulations that use a weight gain retardation of 35 percent result in environmentally attributable health burdens that are approximately similar to the point estimate from WHO experts.

Choice of Parameters

The results of the review informed the choice of parameters used in this study to estimate the revised costs of environmental infections through malnutrition. The literature reviewed shows that diarrhea and other infections cause a weight gain retardation of 20 to 50 percent. Limited research studies how the retardation factor may vary with z-score and age. This study assumes an average loss of 35 percent in terms of weight gain (see table D.4 and step 4 in appendix D) as attributable to infections in early childhood in developing countries. This percentage is used in studies in Ghana and Pakistan to compute counterfactual underweight z-scores and prevalence rates in order to estimate deaths in children under five years of age attributable to infections and malnutrition (presented in chapter 5).

For estimating the effect of stunting on educational performance, this study assumes an average loss of 25 percent in height gain from diarrheal infections in early childhood. The literature reviewed in this appendix suggests that the effect on stunting is somewhat less than on weight gain (see table D.11 and step 10 in appendix D).

Notes

1   Breastfeeding not only diminishes transmission by various mechanisms but also substantially supports energy balance during disease bouts (Brown 2003).

2   Briend and others (1989) observed that the effects of dysentery did not become visible until three months had elapsed.

3   World Health Assembly resolution 54.19, passed in May 2001.

4   Recent expert opinion (Prüss-Üstün and Corvalán 2006) suggests that 50 percent of the health burden attributable to maternal and childhood underweight results from rampant diarrheal disease.

TABLE A.1
Cohort Follow-up Studies Relating Infectious Disease and Nutritional Status of Children in Developing Countries

Country (study)	Setting	Number of Children	Age at Entry	Follow-up Period	Exposure Data Collection	Anthropometric Data Collection	Analysis Strategy	Control Variables (ok if blank)	Notes (ok if blank)	Effect of Disease on Malnutrition
Nigeria (Morley, Bicknell, and Woodland 1968)	Rural	104	6–12 months	Up to 4 years	Symptom of diarrhea accepted only when a clinical examination confirmed diagnosis and illness was considered of sufficient severity to warrant specific treatment	Monthly weights	A child whose weight fell below the 10% percentile at 6, 9, or 12 months) was selected to group A (the under-nourished group); a child who remained between the 50th and 75th percentile (at 9 or 12 months) was selected to group B.	Age	In groups A and B, 8 and 3 children died, respectively.	Group A was characterized by lower birthweights, diarrheal bouts before 6 months, and measles and pertussis bouts at 6–12 months; marked distinction between the groups in weight remained until children were 5 years old.
Guatemala (Martorell, Habicht, and others 1975; Martorell, Yarbrough, and others 1975)	Rural	716	0–7 months	23 months	Retrospective disease reporting every 2 weeks	Every 6 months	Used low versus high diarrhea groups; Used linear regression.	Age, sex, dietary supplement	Children were subject to supplemental nutrition.	Found on average 6.3% (3.5 centimeter) length and 11.0% (1.5 kilogram) weight relative difference; Martorell, Yarbrough, and others' (1975) regression techniques revealed dose-response and 10% of growth deficits would be explainable by diarrhea.
Gambia (Rowland, Cole, and Whitehead 1977)	Rural	152	3 months to less than 3 years	On average, 80 weeks	Self-reported period prevalence (percentage of sick days) of disease between clinic visits (3–10 weeks); diarrhea period prevalence 14%	During clinic visits (3- to 10-week intervals)	Period prevalence was regressed with height and weight gain.	Age, season	Treatment was encouraged and facilitated.	Mean difference between 100% and 0% period prevalence of diarrhea in height and weight gain was –4.2 millimeters and –746 grams per month; regression techniques revealed dose-response.

Country (study)	Setting	Number of Children	Age at Entry	Follow-up Period	Exposure Data Collection	Anthropometric Data Collection	Analysis Strategy	Control Variables (ok if blank)	Notes (ok if blank)	Effect of Disease on Malnutrition
Mexico (Condon-Paoloni and others 1977)	Rural	276	0 months	3 years	Retrospective probing of disease by a health worker every second week to establish period prevalence	Monthly, but weight measured every 2 weeks	Mean growth in upper and lower diarrhea quartiles were compared in a factorial design.	Age, sex, "growth failure"	19 children who developed clinical protein calorie malnutrition were excluded	Diarrhea was the only infection associated with decreased weight gain, and high-diarrhea group attained 95% of the weight of the low-diarrhea group; height was not affected by diarrhea.
St. Lucia, West Indies (Henry 1981)	Rural	229	Less than 6 months	up to 18 months	Examination of stools every 6 months for *Ascaris* and *Trichuris*; mothers instructed to fill out diarrhea cards	Monthly	Quasi-experimental community trial; testing took place in a factorial design of the effectiveness of household water connections and water-seal latrines to prevent helminths.	Comparability by design for confounders: food intake, socioeconomic conditions, general household conditions	*Ascaris* infected children were actively treated with piperazine phosphate.	In the nonintervention village, children remained shorter and weighed less and had helminths, diarrhea, and other water-washed disease more often than in the village, where both household connection and sanitation were provided; when interventions were introduced they led to catch-up growth.
Taiwan, China (Baumgartner and Pollitt 1983)	Rural	118	0 months	More than 6 years	Daily collection by nurses to record frequency, duration, and total days per symptom	Monthly or bimonthly	Twelve disease categories were regressed with growth.	Nutritional treatment, treatment timing, sex	Nutritional supplementation (800 kilocalories) was provided in a placebo-controlled trial.	Negative effect of diarrhea on growth rate was the most important finding, though it explained only 3% of variance; regression techniques revealed dose-response.
Northeastern Brazil (Guerrant and others 1983)	Small town plus rural adjacent areas with varying	297 study participants, of which 53 subjects were used to study effects on nutrition	0–6 years	3 months	Daily diary by mother or legal guardian plus weekly visits by a surveillance team recording clinical symptoms to establish period prevalence;	Quarterly (height); weekly (weight)	Regression analysis of 482 child-periods of 3 months were regressed against the number of diarrhea episodes	Age, breast-feeding, growth status before the period	Antidiarrheal treatments (iodochlorhydroxyquin and kaolin pectin) were given.	Multiple different pathogens were found in 50% of episodes, and height faltering attributable to episodes was especially pronounced (41%) during the second year of life; Breastfeeding found to be highly

(continued)

TABLE A.1
Cohort Follow-up Studies Relating Infectious Disease and Nutritional Status of Children in Developing Countries (continued)

Country (study)	Setting	Number of Children	Age at Entry	Follow-up Period	Exposure Data Collection	Anthropometric Data Collection	Analysis Strategy	Control Variables (ok if blank)	Notes (ok if blank)	Effect of Disease on Malnutrition
	levels of environmental health services				collection of stool specimens by health worker					protective, especially in first 6 weeks. Regression techniques revealed dose-response.
Bangladesh (Black, Brown, and Becker 1984)	Rural	157	6–48 months	12 months	Period prevalence of diarrhea within 60-day periods as actively established by a health worker who visited homes every 2 days	Monthly	Growth in zero-period prevalence periods was compared with growth in nonzero periods.	Age, weight and length at base-line	All cases of diarrhea were medically treated.	Children with no diarrhea gain on average 0.42 centimeter more in length per year compared with children suffering from diarrhea; diarrhea also adversely affected weight gain; invasive *Shigella* and enterotoxigenic *E. coli* might especially cause irreversible growth faltering; regression techniques revealed dose-response.
Bangladesh (Henry and others 1987)	Rural villages	On average, 300 children	5–24 months	6–19 months	Period (60-day) prevalence and attack rates based on weekly visits by health assistants	Five anthropometric surveys every 6 months	Mean duration of watery diarrhea and dysentery in different anthropometric categories were tabulated.	Age		No association of watery diarrhea with growth faltering was noted, but stunting was associated with dysentery.
Bangladesh (Bairagi and others 1987)	12 rural villages in Matlab	Approximately 1,000	1–4 years	9 months	Weekly visits by field workers to probe (7-day recall) diarrhea episodes and duration; no distinction between watery diarrhea and dysentery	Bimonthly	Multiple regression analysis was used to investigate diarrhea-nutrition relationship.	Age, dwelling floor space (proxy for socioeconomic status), mother's education	Several observational epidemiological studies led by International Centre for Diarrhoeal Disease Research were carried out in Matlab region.	Those with greater than 10% diarrhea period prevalence had 71% (weight) and 75% (height) growth velocity throughout the follow-up period, compared with those who did not have diarrhea; catch-up growth among those who had less than 10% diarrhea period prevalence prevented growth faltering.

Country (study)	Setting	Number of Children	Age at Entry	Follow-up Period	Exposure Data Collection	Anthropometric Data Collection	Analysis Strategy	Control Variables (ok if blank)	Notes (ok if blank)	Effect of Disease on Malnutrition
Sudan (Zumrawi, Dimond, and Waterlow 1987)	Urban	439, but falling to 262 due to drop-outs and families moving away	0 months	1 year	Morbidity probed at visits every 2 weeks visits on the basis of 14-day recall; period prevalence 6%	Every 2 weeks	Number of days ill were regressed with weight gain; weight loss due to infections were externally (WHO reference) and internally between quartiles assessed	Age		Weight gain deficits from 1 day of diarrhea and cough or cold was 32 and 16 grams, respectively; External comparison revealed that almost all weight gain deficit was due to infections, though internal comparison revealed little weight gain deficit attributable to infections; regression techniques revealed dose-response.
Gambia (Rowland, Rowland, and Cole 1988)	Urban	126	0 months	24 months	Number of illness days (10 categories) probed 1 month retrospectively during every clinic visit to establish period prevalence	Monthly	Illness growth periods were compared with no-illness periods.	Age, season	Study infants had ready access to a pediatrician whenever they were ill.	Combined effect of acute lower respiratory infection (-320) and diarrhea (-610) was 930 grams, but there was no effect on height; acute lower respiratory infection impact varied by season; effects were exclusively found among non-breastfed babies; regression techniques revealed dose-response.
Bangladesh (Briend and others 1989)	Rural	On average 230[a]	6-35 months	42 months 'fourteen quarterly rounds'	Height and weight measured quarterly; diarrhea morbidity collected by means of weekly recall by the children's mothers	Quarterly (varying)	Growth rates of diarrhea-free periods were compared with growth rates of periods with varying prevalence of diarrhea, with attention to timing of disease within studied periods.		Effective case management with oral rehydration therapy was organized within the study; 40.7% of periods excluded because of missing data	Only transient changes in growth were observed, and were caught-up later, though dysenteric patients were observed to experience stunting with a 3-month lag.

(continued)

TABLE A.1
Cohort Follow-up Studies Relating Infectious Disease and Nutritional Status of Children in Developing Countries *(continued)*

Country (study)	Setting	Number of Children	Age at Entry	Follow-up Period	Exposure Data Collection	Anthropometric Data Collection	Analysis Strategy	Control Variables (ok if blank)	Notes (ok if blank)	Effect of Disease on Malnutrition
Colombia (Lutter and others 1989)	Urban setting with at least half of siblings under 5 who were less than 85% weight for age.	148 (control), 140 (supplement)	0 months	36 months	Semimonthly visits to establish incidence and duration of diarrhea, vomiting, and respiratory and skin infections to establish period prevalence; diarrhea prevalence 4.5–12.2% according to age	Semimonthly	Linear regression and 2-way analysis of variance were used to model either increment of weight and height since birth or absolute height and weight until 36 months.	Age, treatment group, size at birth	Throughout the study, families received free obstetric and pediatric care, including medicine; the supplement group received 600 kilocalories and 30 grams of protein.	Each day ill with diarrhea is associated with a reduction 0.03 cm (p < 0.001) in attained length at age 36 mo; food supplements effectively mitigated growth faltering (4.9 centimeters by age 36 months), especially among children who experienced greater than 10% diarrhea period prevalence; the difference between this group and the WHO (World Health Organization) standard was 13 centimeters at 36 months (that is, the treatment group experienced growth faltering of about 8 centimeters); regression techniques revealed dose-response.
Brazil (Schorling and others 1990)	Urban slum in northeastern Brazil, where an exceptionally high rate of endemic diarrhea persists	175	0–5 years	28 months	Three times a week; during third year, extensive stool specimens collected from acute, persistent, and control children; specimens subject to intensive microbiological testing	Monthly	Descriptive time-series analysis was made.	Age		Authors noted that the greatest contribution to the declines came from a 50% increase in the prevalence of persistent diarrhea and that this decline closely followed a decrease in the prevalence of severe malnutrition (weight for age) (p = 0.01); control children also had high rates (37%) of diarrheagenic agents.

Country (study)	Setting	Number of Children	Age at Entry	Follow-up Period	Exposure Data Collection	Anthropometric Data Collection	Analysis Strategy	Control Variables (ok if blank)	Notes (ok if blank)	Effect of Disease on Malnutrition
Bangladesh (Becker, Black, and Brown 1991)	Rural villages in Matlab area	66	5–18 months	Up to 14 months	Home visits every second day by health worker, who recorded diarrhea, vomiting, nasal discharge, cough, anorexia, and fever	Monthly	Three different random effects models were used, with morbidity period prevalence fitted as independent variables.	Energy intake, age	Matlab is a special observation region, where many observational epidemiological studies are carried out by the International Centre for Diarrhoeal Disease Research.	If diarrhea and fever were reduced to the level of a U.S. study children would gain 226 grams per month more; regression techniques revealed dose-response.
Brazil (Victora and others 1990)	Relatively wealthy urban setting	5,914	0 months	36–52 months	During two surveys every second year, mothers probed about hospitalizations caused by, for example, pneumonia and diarrhea; validation done by reviewing records of 120 hospitalizations	Every second year	Comparisons were made of mean standard z-scores among hospitalized and nonhospitalized children.	Age, sex, income level, birthweight		Effect of early infancy hospitalizations strongly predicted growth faltering (at 36–52 months), though pneumonia—unlike diarrhea—was statistically significantly linked with stunting (for example, early infanthood diarrhea and pneumonia hospitalizations decreased mean height-for-age z-score by about 50% and 33%, respectively).
Zimbabwe (Moy and others 1994)	Rural impoverished setting where caloric intake was 65% of recommended	204	0–12 months	22 months	Weekly maternal recall to establish period prevalence	Monthly	Comparisons of growth rates between diarrhea-free periods and periods with varying period prevalence of diarrhea, with attention paid to timing of disease within studied		Children were exclusively breastfed until 3 months; all children were fully immunized, and were mothers trained to treat children with oral rehydration	Authors found very little difference in growth rates with the most diarrhea and with least; study showed that a diarrheal episode led to 51 grams and 0.18 centimeters of growth loss, which could not explain the significant growth faltering in the in the cohort;

(continued)

Cohort Follow-up Studies Relating Infectious Disease and Nutritional Status of Children in Developing Countries *(continued)*

Country (study)	Setting	Number of Children	Age at Entry	Follow-up Period	Exposure Data Collection	Anthropometric Data Collection	Analysis Strategy	Control Variables (ok if blank)	Notes (ok if blank)	Effect of Disease on Malnutrition
							periods; in addition, growth rates were compared between		supplements. high (more than 9) and low (fewer than 4) occurrence of diarrhea in 22 months.	timing of diarrhea within the 3-month periods suggested that growth faltering was transient, though with respect to linear growth it was not clear cut.
Philippines (Adair and others 1993)	Urban and rural areas around and within Cebu	2,186 (+375) infants	0 months	24 months	Bimonthly maternal recall (prior week) of acute respiratory infection and diarrhea episodes	Bimonthly	Lagged weight gain was regressed against morbidity; an instrumental variables technique was used as explanatory variables were correlated.	Breast-feeding and energy intake from other sources, preventive health care, maternal weight, season of measurement		Recent diarrhea had a substantial significant negative effect on weight of children after 6 months of life, whereas acute respiratory infections significantly negatively affected growth up to 6 months of life. Regression coefficients: approximately 250-400 grams lagged or immediate effects of diarrhea or acute respiratory infections in either 0-6-month or 6-24-month period; results are suggestive of lasting irreversible effects; breastfeeding was strongly protective.
Sudan (Harrison, Brush, and Zumrawi 1993)	Poor peri-urban Khartoum	120 mother-infant pairs	0 months	12 months	Biweekly recall	Monthly	Comparison was made of mean duration of illness and growth of children born to housewives and nonhousewives			Housewives' children were on average 3.8 centimeters longer at 12 months although they were shorter at birth; on average illness duration at 2-12 months was up to 5 times onger among children born to working mothers despite the fact that such children tended I to have

Country (study)	Setting	Number of Children	Age at Entry	Follow-up Period	Exposure Data Collection	Anthropometric Data Collection	Analysis Strategy	Control Variables (ok if blank)	Notes (ok if blank)	Effect of Disease on Malnutrition
										less access to water supply, sewage systems, and general conditions of hygiene.
Honduras (Cohen and others 1995)	Low-income communities in San Pedro Sula, where environmental sanitation was poor	141 mother-child pairs participating in a randomized trial of 3 treatment groups	0 months	12 months	Morbidity data (upper respiratory infection, fever, diarrhea) collected at less than 4 months during weekly visits (7-day recall); mothers kept stool frequency and consistency records to establish period prevalence (about 3.2% for diarrhea); at greater than 6 months, monthly visits probed 30-day recall.	Monthly between 7 and 12 months; before that, either weekly or more irregularly at less than 16 weeks	Comparison was made of growth velocities between treatment groups and U.S. standards at 0-4 months, 4-6 months, 6-12 months.	Sex; birth-weight; maternal height, age, and education; group, income, potable water;, intervention group, breast milk energy at 4 months, income, maternal BMI at 4 months	Willingness to breastfeed for 26 weeks was an inclusion criterion, and during weekly visits this habit was encouraged; hygienic food was provided during post-weaning period for those (2 arms) who discontinued exclusive breastfeeding, according to the intervention protocol.	There was an initial rapid catch-up compared with the U.S. data at less than 3 months; after 6 months and especially at greater than 9 months, rapid growth-faltering; intrauterine growth restriction was a strong predictor of stunting; fever, but not diarrhea, predicted growth faltering between 6 and 12 months; the low morbidity rates and lack of association of diarrhea with growth may be attributable to the study's emphasis on breastfeeding.
Sudan (Brush, Harrison, and Waterlow 1997)	Peri-urban Khartoum	167	0 months	12 months	Biweekly mother's recall of the duration of illness	Monthly	Infancy was divided into periods of greater than 6 months and less than 6 months, and biweekly period prevalence was regressed with growth parameters.	Mother's socioeconomic status and income, age of weaning, and disease duration at under 6 months		Duration of illness in the first 6 months was strongly inversely related to length and weight growth at 12 months.

(continued)

TABLE A.1
Cohort Follow-up Studies Relating Infectious Disease and Nutritional Status of Children in Developing Countries *(continued)*

Country (study)	Setting	Number of Children	Age at Entry	Follow-up Period	Exposure Data Collection	Anthropometric Data Collection	Analysis Strategy	Control Variables (ok if blank)	Notes (ok if blank)	Effect of Disease on Malnutrition
Guinea-Bissau (Molbak and others 1997)	Peri-urban district of Bandim II	1,064 followed for 1,441 child-years	0–? months (462 recruited at birth)	36 months	Biweekly visits by field workers; stool specimens collected the same day if child was sick	Approximately 3 months	Growth faltering was studied in different age groups (less than 1 year, 1–2 years) with three different lag periods and a common estimate.	Preinfection weight, sex, age, season, and age at infection		Long-lasting growth-faltering effect was observed in both weight and height velocities, effects being most pronounced at under 1 year. At less than 1 year, cryptosporidiosis lowered weight by 392 and 294 grams, among boys and girls respectively; respective results for height were −4.7 and −8.3 millimeters.
Indonesia (Kolsteren, Kusin, and Kardjati 1997)	Rural Indonesia, where childhood infections are relatively rare	141	0–24 months	36 months	Weekly visits by health worker; ill children revisited after 3 days to establish period prevalence of acute respiratory infection, diarrhea, and fever	Monthly	Period prevalence of acute respiratory infections and diarrhea was associated with growth velocity in 4-week periods.	Duration of a specific disease episode, sex, age	Sick children were referred to the project clinic; virtually all were breastfed until 12 months.	Weight indexes were significantly affected, especially by acute respiratory infections after 6 months of age, but no effect on linear growth was observed; duration of diarrhea, acute respiratory infections, and fever did not contribute to length velocity, but weight velocity was adversely affected by acute respiratory infection duration.
Peru (Checkley and others 1997)	Peri-urban shanty-town	207	0–3 months	23 months	Stool specimens collected weekly irrespective of symptoms; diarrhea symptoms probed biweekly by a health worker	Monthly	Comparison was made of growth velocity in 1-month growth periods among those infected by *Cryptosporidium* (first time infected	Sex, weight, and stunting indicator at the start of period; interaction of *C. parvum* and diarrhea; presence of *G. lamblia*; seasonality		On average, symptomatic and asymptomatic children gained 342 and 162 grams less after the first *Cryptosporidium* infection; because asymptomatic infections are twice as common, they might have a bigger public health effect through malnutrition.

Country (study)	Setting	Number of Children	Age at Entry	Follow-up Period	Exposure Data Collection	Anthropometric Data Collection	Analysis Strategy	Control Variables (ok if blank)	Notes (ok if blank)	Effect of Disease on Malnutrition
Brazil (Steiner and others 1998)	Shanty-town in north-eastern Brazil	186	0 months	5 years	Stool samples collected periodically from healthy children and during each episode of diarrhea	Quarterly	Growth of E. coli (EAggEC) positive children with persistent and acute diarrhea and children without diarrhea were compared with growth of children who were not carrying enteric pathogens.	Presence of pathogens other than E coli EAggEC	Cohort was well described.	Regardless of presence of diarrhea, EAggEC positive children showed significantly increased growth faltering (more than 1.5 drop in Z-score) 3 months after infection. EAggEC children without diarrhea (n = 6) showed intestinal inflammation and also had growth faltering (−0.64 in z-score; p = 0.064); EAggEC remains dramatically underreported because showing its presence requires special techniques. 15–31% of asymptomatic children in Brazil and 6.5–9.9% in India are carriers.
							included and first month after infection covered) and those not infected.			
Peru (Checkley and others 1998)	Peri-urban Lima	185	0–3 months	24 months	Weekly collection of stool specimens to diagnose C. parvum and biweekly diagnosis of diarrhea to establish period prevalence	Monthly measurements by health workers	Natural cubic-B-spline was used to model growth, with special attention paid to 63 children who became infected with C. parvum during the study.	Age prior stunting, food expenses, number of diarrheal episodes, persistent diarrhea, water source and quality		C. parvum when acquired very early after birth–and especially if child is stunted prior to infection–likely leads to irreversible stunting; C. parvum contributes to about 1 centimeter height loss after 1 year of infection.

(continued)

TABLE A.1
Cohort Follow-up Studies Relating Infectious Disease and Nutritional Status of Children in Developing Countries *(continued)*

Country (study)	Setting	Number of Children	Age at Entry	Follow-up Period	Exposure Data Collection	Anthropometric Data Collection	Analysis Strategy	Control Variables (ok if blank)	Notes (ok if blank)	Effect of Disease on Malnutrition
Brazil (Lima and others 2000)	Shanty-town in north-eastern Brazil	39–42 (first-ever persistent diarrheas found among 179 children)	0 months	Up to 48 months	Examination 3 times weekly by a nurse; extensive sampling of diarrheal pathogens from those who developed persistent diarrhea and matched controls	Quarterly	Growth at 3 months before and after first-ever persistent diarrhea episode was compared with that of children who did not develop persistent diarrhea.	Age, sex, feeding practices		Statistically significant short-term growth faltering (weight for age and height) was found, but there was no effect on height for age; *Cryptosporidium, Giardia,* enteric adenoviruses, and enterotoxigenic *E. coli* were significantly associated with persistent diarrhea; even partial breastfeeding was found to be protective against diarrheal disease; persistent diarrhea was strongly associated with lack of toilet (pit or flush).
Bangladesh (Alam and others 2000)	Rural Matlab region	412	6–48 months	6–48 months	Home visits by trained health workers every 4 days; and watery diarrhea and dysentery episodes recorded to establish period prevalence (approximately 10%)	Monthly	Weight and height gains in no infection, watery diarrhea, and dysentery groups were compared by using 3-month or 1-year periods.	Age, sex, parental education, household income, land ownership, baseline nutritional status, diarrhea period prevalence	Matlab is a special observation region, where many observational epidemiological studies are carried out by the International Centre for Diarrhoeal Disease Research.	Strong evidence was found that watery diarrhea, but especially dysentery, causes growth faltering in the short and long terms. Watery diarrhea and dysentery resulted in 300 grams (0.68 centimeters) and 516 grams (1.34 centimeters) of statistically significant growth faltering, respectively, in a 1-year period.

Country (study)	Setting	Number of Children	Age at Entry	Follow-up Period	Exposure Data Collection	Anthropometric Data Collection	Analysis Strategy	Control Variables (ok if blank)	Notes (ok if blank)	Effect of Disease on Malnutrition
Pakistan (Fikree, Rahbar, and Berendes 2000)	Urban squatter settlements of Karachi	994 pregnant women initially recruited; 738 newborns recruited for follow-up	n.a.	Children visited at one, three, six, nine, twelve, eighteen and twenty-four months	Previous month recall (acute lower respiratory infection) or 2 weeks (diarrhea) during visits; for child to be classified as having diarrhea he or she had to have diarrhea in at least 2 of 3 contact times (that is, exposure relied on the number of episodes)	1-, 3-, 6-, 9-, 12-, and 24-month visits	Logistic regression technique used to estimate relative risk of becoming malnourished when child was exposed to infections	Socioeconomic indicators (income, house, water supply); maternal indicators (education, weight, parity, first child); feeding practice (breast or nonhuman milk)		Intrauterine growth restriction was the strongest predictor of stunting and wasting; water supply and diarrheal episodes predicted stunting, but risk factors for wasting were limited only to socioeconomic status. Stunting rose from 27.5% at 3 months to 67.9% at 18 months, which the authors noted could be due to feeding practices and exposure to a more contaminated environment than found in Western countries; breast-feeding was protective.
Brazil (Moore and others 2001)	Shanty-town in north-eastern Brazil	119	0 months	3 years, 9 months	Home visits 2 or 3 times weekly to establish clinical period prevalence of diarrhea and stool samples to establish enteric pathogens, ova, and parasites	Quarterly	Repeat measure modeling was used to explain growth faltering caused by early childhood diarrhea and intestinal helminths.	Nutritional status in infancy, other parasitic infections, diarrhea in later childhood, socioeconomic status	Results represent a best-case scenario as this shantytown has been subject to substantial nutritional interventions; Ascaris infections were treated.	Study found early childhood diarrhea and intestinal helminths were independently associated with profound and lasting linear growth faltering. By age 7, an average child experiencing 9.1 diarrhea episodes before age 2 had a 3.6-centimeter growth deficit; early helminth infection added 4.6 centimeters to this deficit.
Guinea-Bissau (Valentiner-Branth and others 2001)	Peri-urban setting	107	0–2 years	6.6 months (median)	Morbidity recorded during weekly visits; period prevalence 10.3%	Monthly (knee-to-heel length) and weight, total length quarterly	Natural growth history of children with persistent diarrhea was studied.		Children were examined by a physician and offered fluid therapy and antibiotics, according to WHO guidelines; breast feeding was encouraged; half of children received nutritional interventions.	Authors found that the negative effect of persistent diarrhea on linear growth was sustained during the follow-up period, but some catch-up in weight gain occurred in the treatment group.

(continued)

137

TABLE A.1
Cohort Follow-up Studies Relating Infectious Disease and Nutritional Status of Children in Developing Countries *(continued)*

Country (study)	Setting	Number of Children	Age at Entry	Follow-up Period	Exposure Data Collection	Anthropometric Data Collection	Analysis Strategy	Control Variables (ok if blank)	Notes (ok if blank)	Effect of Disease on Malnutrition
Egypt (Wierzba and others 2001)	Peri-urban Alexandria, where malnutrition rates remain below 20%	143	0–24	12–22 months	Twice-weekly home visits to establish number of diarrhea episodes	Approximately quarterly	Study analyzed odds ratio to fall below –2 standard deviations greater than or equal to 1 episode versus none during interval; additional special emphasis was placed on 6-month periods where diarrhea occurs in the first half, to study catch-up growth in a disease-free environment.	Age, socioeconomic status, mother's education, days of exclusive breastfeeding, presence of sanitary latrine	Oral rehydration therapy was provided, and referrals for hospital care were made when needed; exclusive breast-feeding was ascertained during visits.	There was a –0.16 ($p < 0.05$) reduction of the standard z-score; when periods extended to 6 months and diarrhea occurred during the first half of the period, the association remained negative in models but was not statistically significant; authors noted that catch-up may compensate growth faltering, but those with diarrhea in the first half when compared to nondiseased individuals had slower growth velocity ($p > 0.05$).
Peru (Checkley and others 2003)	Peri-urban Lima	224	0 months	35 months	Active surveillance of diarrhea to establish clinical period prevalence and stool samples to establish pathogens	Monthly	CAR1 error term modeling was used based on 156,436 child-days with 3,335 days of diarrhea (2.1%).	Age, sex, breast-feeding, water supply and sanitation maternal stature	Study does not specifically mention, but as children were attended daily, case management must have been included.	Diarrheal period prevalence in the cohort explained 2–27 % of the growth retardation, and the average child was 2.5 centimeters shorter than international peers; diarrhea, especially before age 6 months, predisposes children to irreversible stunting; with extensive resolution of collected data and elaborate modeling, this study showed decelerations on linear growth later in life as a result of early infancy exposure to diarrhea; a 10 cm difference in maternal stature corresponded to a 1.5 cm height deficit at 24 months

Country (study)	Setting	Number of Children	Age at Entry	Follow-up Period	Exposure Data Collection	Anthropometric Data Collection	Analysis Strategy	Control Variables (ok if blank)	Notes (ok if blank)	Effect of Disease on Malnutrition
Malawi (Maleta and others 2003)	Rural (covered 95% of deliveries in the catchment area)	767	0 months	36 months	Data on infant feeding and morbidity collected monthly until 18 months; no morbidity data collected thereafter	Monthly until 18 months; thereafter quarterly until 36 months	Odds ratio of falling below severe (–3 standard deviations) underweight and stunting and moderate (?2 standard deviations) wasting during follow-up, using greater than or equal to 2.4 versus less than 2.4 disease episodes per month in infancy as the explanatory variable.	Average size at 1–3 months, distance to health center, gestation less than 37 weeks, maternal HIV, delivery by trained assistant or not, sex	High attendance rates (128 died and only 62 dropped) were a feature throughout the study; during the first prenatal visit mothers were offered medical examinations and laboratory measurements.	Infancy disease episodes were strongest determinants of growth faltering, followed by intrauterine growth restriction (using a proxy measure); disease episodes remained strong predictors of underweight and wasting, but adjustment with intrauterine growth restriction (which remained a highly significant predictor of stunting) dropped infancy disease episodes from 3.4 to 1.1 (adjustment with itself).
Tanzania (Villamor and others 2004)	Urban Dar es Salaam hospital (children admitted recruited to receive a series of A-vitamin shots)	524	6–60 months	12 months	Morbidity episodes (diarrhea, on average 3.4 episodes per child per year) established during monthly visits using pictorial diary; blood samples from all children at hospital; blood collected from subsamples later	Monthly	Mixed-effect models were estimated to see how much HIV infection causes growth faltering; adjusted various morbidity effects on growth faltering were also estimated.	Age, index episodes, HIV, maternal education, treatment arm, anemia		HIV was strongly associated—especially in the 6- to 11-month period with growth faltering (1.3 centimeters and 1.3 kilograms), whereas diarrhea, though it decreased length in the 6-11-month period by 1.4-2 centimeters, did not show linear dose-response; dysentery appeared to be related to lower weight gain (0.32 kilograms) in children in the 12-23-month group.

(continued)

TABLE A.1

Cohort Follow-up Studies Relating Infectious Disease and Nutritional Status of Children in Developing Countries *(continued)*

Country (study)	Setting	Number of Children	Age at Entry	Follow-up Period	Exposure Data Collection	Anthropometric Data Collection	Analysis Strategy	Control Variables (ok if blank)	Notes (ok if blank)	Effect of Disease on Malnutrition
Brazil (Prado and others 2005)	Urban	597	6–45 months	16 months	Stool samples collected once during the first year, but diarrhea symptoms probed twice a week to establish period prevalence	Every 6 months	Variance and covariance were analyzed.	Sex, age, house floods during rain, presence of rubbish near house, duration of interval between measurements, prevalence of diarrhea during that interval and intensity of infection, *Ascaris lumbricoides* and *Trichuris trichiura* baseline anthropometry	All children were treated to eradicate *Giardia lamblia*, *Ascaris lumbricoides*, and *Trichuris trichiura* during the first year of study.	Both symptomatic and asymptomatic *Giardia* infections had a lasting effect on linear growth; investigators were able to exclude reverse causality (that is, the mere susceptibility of the malnourished to *Giardia* infection as the explanation for the short stature of those infected) by showing the effect on stunting with a lag.
Colombia (Alvarado and others 2005)	Remote rural Afro-Colombian village (Guapi)	133	5–7 months	12 months	Morbidity recorded by mothers recorded daily (diarrhea, cough, fever, or other nonspecific symptoms); data collected by a health worker weekly	Approximate 3-month intervals	Hierarchical linear model was used to model time-varying and fixed variables on weight and height gain.	Initial status, gender, preterm or term delivery, maternal height, mother's education, child's age, breastfeeding	Continued breast-feeding (more than 6 months) was widespread at 15 months; women increased breastfeeding during disease episodes, unlike in many other cultures.	Study showed the substantial protective effect of breast-feeding, which diminished with increasing socioeconomic status and which probably mitigated growth faltering caused by diarrhea morbidity; respiratory infections were associated with growth faltering.

Source: Compiled by World Bank team.

a. Not a fixed cohort, because surveys were made in an "open cohort."

APPENDIX   B

Review of Studies on Nutritional Status and Education

THIS APPENDIX REVIEWS RECENT LITERATURE on the effects of low height for age in childhood on subsequent educational outcomes.[1] One group of studies investigates the effects of malnutrition on cognitive function, measured in terms of IQ scores or performance on cognitive tests (Fishman and others 2004). A second group assesses the effect on educational outcomes. This appendix concentrates on the latter.

Several longitudinal studies have recently become available that investigate the effect of early childhood malnutrition on educational outcomes (table B.1). The majority of longitudinal studies use stunting (low height-for-age z-score, or HAZ) as a measure of nutritional status, because HAZ is considered an indicator of chronic nutritional deficit. These studies use data that follow children over time, usually from birth well into primary or secondary schooling, and contain various anthropometric, schooling, and socioeconomic status indicators. Such longitudinal data have important advantages in terms of investigating causal relationships between malnutrition and educational outcomes compared with the cross-sectional data often used in earlier studies.

Grantham-McGregor and others (2007) present an analysis of longitudinal data of more than 2,000 male youths in Brazil. They find that stunting (HAZ < –2) in early childhood is associated with a loss of 0.9 years of grade attainment by age 18. Using the stratified data in their study, Grantham-McGregor and others (2007)

TABLE B.1
Studies of the Effects of Malnutrition on Educational Outcomes

Source	Country	Data	Children	Indicator	Impact on Education
School grade attainment					
Grantham-McGregor and others 2007	Brazil	Longitudinal, 1982–2001	2,041 males	HAZ < –2; HAZ –2 to –1 versus > –1	0.91 lower grade attainment; 0.44 lower grade attainment[a]
Maluccio, Hoddinott, and Behrman 2006	Guatemala (rural)	Longitudinal, 1969–2004	1,469	Nutrition supplement	1.0 higher grade attainment (women)
Daniels and Adair 2004	Philippines (Cebu)	Longitudinal, 1983–2002	1,970	HAZ < –3; HAZ –3 to –2; HAZ –2	0.80 lower grade attainment; 0.33 lower grade attainment; 0.12 to –1 versus > –1 lower grade attainment
Alderman and others 2006	Uganda (rural)	Longitudinal, 1983–2000	570	HAZ	0.68 lower grade attainment per 1 HAZ decline
Learning productivity (grade equivalents)					
Behrman and others 2006	Guatemala (rural)	Longitudinal, 1969–2004	1,448	HAZ	0.84 school year equivalent per 1 HAZ
Glewwe, Jacoby, and King 2001	Philippines (Cebu)	Longitudinal, 1983–95	1,016	HAZ; HAZ < –2.5	0.8 school year equivalent per 1 HAZ; 1.8 school year equivalent per 1 HAZ
School dropout and grade repetition					
Maluccio, Hoddinott, and Behrman 2006	Guatemala (rural)	Longitudinal, 1969–2004	1,469	Nutrition supplement	0.1 grade per year faster school progression (women)
Walker and others 2005	Jamaica (Kingston)	Longitudinal, 1986–2003	167	HAZ < –2 versus HAZ > –1	100% higher school dropout rate by age 17–18

				HAZ < -3; HAZ -3 to -2; HAZ -2 to -1	
Daniels and Adair 2004	Philippines (Cebu)	Longitudinal, 1983–2002	1,970	HAZ < -3; HAZ -3 to -2; HAZ -2 to -1	18% grade repetition; 11% grade repetition; 5% grade repetition[b]
Primary school enrollment					
Glewwe and Jacoby 1995	Ghana	Cross-sectional 1988–89	1,757	HAZ	0.34 year delayed school enrollment per 1-unit decrease in HAZ
Alderman and others 2001	Pakistan (rural)	Longitudinal, 1986–91	534	0.5 HAZ increase	18% increase in girls' enrollment; 4% increase in boys' enrollment
Glewwe, Jacoby, and King 2001	Philippines (Cebu)	Longitudinal 1983–95	1,016	HAZ	2 months delayed enrollment per 1-unit HAZ decline
Alderman and others 2006	Uganda (rural)	Longitudinal, 1983–2000	570	HAZ	0.4 year delayed enrollment per 1-unit HAZ decline
Effect of diarrhea					
Lorntz and others 2006	Brazil (shantytown)	Longitudinal 1993–2001	77	Number of diarrheal episodes (0–2 years)	0.7 month delayed school enrollment per episode of diarrhea

Source: Compiled by World Bank team.
a. Estimated here from table 5 in Grantham-McGregor and others (2007).
b. Repetition rates are relative to less than −1 standard deviation; estimated here from odds ratios and data presented in Daniels and Adair (2004).

estimate that mild stunting (HAZ –1 to –2) is associated with a loss of 0.4 years of schooling.

Alderman and others (2001), using longitudinal data relating to 560 children in rural Zimbabwe, find that a decline in HAZ of 1 is associated with a loss of 0.68 years of grade attainment and 0.4 years of delayed primary school enrollment. *Grade attainment* is defined as completed grades in primary and secondary school.

Behrman and others (2006), using longitudinal data of 1,448 children followed into adulthood in rural Guatemala, assess the effect of preschool, schooling, and postschooling experiences on adult cognitive skills (reading comprehension and nonverbal skills). They find that preschool nonstunting (HAZ ≥ –2) increases reading comprehension test scores by the same amount as 4.2 additional years of schooling attainment. Using sample and subsample mean values of HAZ reported in the study, an improvement in HAZ of 1 has an effect on reading comprehension equivalent to nearly one additional year of schooling attainment. Stunting is also found to have a profound negative effect on nonverbal test scores. To the extent that nonverbal cognitive skills may affect learning productivity per year of schooling, this estimated effect of a change in HAZ of 1 on schooling attainment is conservative.

Maluccio, Hoddinott, and Behrman (2006) use the same data set to evaluate the effect of early childhood nutrition supplementation in Guatemala. The supplement was found to decrease the prevalence of severe stunting by 50 percent, increase the height of three-year-old children by 2.5 centimeters, and increase school grade attainment by one full year for women. No significant effect on schooling was found for men. The effect for women comes through a reduction in women who never attended school and an increase in women who progressed into secondary school. Women who received supplementation progressed faster through school than women who did not receive supplementation, at a rate of 0.1 grade per year in school.

Walker and others (2005), using longitudinal data of 167 children in Kingston, Jamaica, find that 17- to 18-year-olds, stunted by the age of two (HAZ < –2), had a school dropout rate twice as high as children who were not stunted (HAZ > –1).

Daniels and Adair (2004) use longitudinal data of 1,970 children in the Cebu metropolitan area (urban and rural households) in the Philippines. They estimate odds ratios for completing high school, delayed enrollment, and grade repetition in relation to early childhood stunting. Using the odds ratios and data presented in their article, one can infer that severe stunting in early childhood (HAZ < –3) is associated with 0.80 years of lower grade attainment, that moderate stunting (HAZ –2 to –3) is associated with 0.33 years of lower attainment, and that mild stunting (HAZ –1 to –2) is associated with 0.12 years of lower attainment. Daniels and Adair (2004) also find that severe stunting is associated with repeating a grade at least once (18 percent of severely stunted children) and that 5 percent of mildly stunted children repeat a grade at least once as a result of stunting.

Glewwe, Jacoby, and King (2001), using longitudinal data relating to more than 1,000 children in the Cebu metropolitan area, investigate the effect of stunting on learning productivity per year of schooling through primary school using achievement test scores (sum of math and English). They then convert the effect of stunting on test scores into the extra months of schooling it takes to compensate for the difference in test scores. They find that the effect on test scores of a change in HAZ of 1 is equivalent to 0.8 year of school attendance, and for the more severely stunted children (HAZ < −2.5), it is equivalent to 1.8 years of schooling. Glewwe, Jacoby, and King (2001) also find that a decline in HAZ of 1 delays primary school enrollment by two months and increases the probability of first-grade repetition by 9 percent. Using the baseline data in Daniels and Adair (2004) on frequency of grade repetition by stunting status, they find that a grade repetition of 9 percent is equivalent to a 0.5-month delay in labor force entry for a severely stunted child and a 0.4-month delay for a mildly stunted child.

Alderman and others (2001), using longitudinal data on more than 500 children in rural Pakistan, find that a 0.5 increase in HAZ is associated with an 18 percent increase in girls' school enrollment and a 4 percent increase in boys' school enrollment.

Glewwe and Jacoby (1995) find in a study of children in Ghana, albeit using cross-sectional data, that a decline in HAZ of 1 is associated with a 0.34-month delay in primary school enrollment.

Diarrhea and Education

Lorntz and others (2006), using longitudinal data on children in a shantytown in northern Brazil, estimate the effect of early childhood diarrheal episodes on subsequent timing of primary school enrollment. They find a 0.7-month delay in enrollment per diarrheal episode in the first two years of life. The median number of episodes in the first two years of life was 4.5 per year. On average, this finding corresponds to a two-week prevalence rate of nearly 22 percent, which is the same as reported in the 2003 Ghana Demographic and Health Survey for children under two years of age. Reducing this prevalence rate by half (that is, to 11 percent) would suggest a reduction in delayed enrollment by about three months.

Conclusions

The literature reviewed and presented in table B.1 provides substantial evidence of a significant effect of malnutrition on children's educational performance. The single most important effect is reduced learning productivity during schooling, followed by reduced grade attainment. Delayed primary school enrollment and grade repetition, while important in their own rights, have a much lesser consequence for lifetime income (see appendix D and chapter 5).

The effects of malnutrition on educational performance, as applied to Ghana and Pakistan to estimate the impact on lifetime income, are presented in table 5.6 (see also table D.7 in appendix D). The attributable fraction of malnutrition (stunting) from diarrheal infections associated with water, sanitation, and hygiene is discussed in appendixes A and D. The relation between educational performance and future income is discussed in chapter 5.

Note

1 This appendix is based on a review of several recent longitudinal studies that looked at the effect of low height for age in childhood on educational outcomes. This review was carried out by Bjørn Larsen (consultant) specifically for this report.

New Estimates for Burden of Disease from Water, Sanitation, and Hygiene

THE ESTIMATES IN TABLE C.1 ARE FROM the new study by the World Health Organization (Fewtrell and others 2007). The study incorporates the effects (through malnutrition) of water, sanitation, and hygiene on mortality and disease burden.

TABLE C.1

Burden of Disease (in DALYs) in Children under Five Years Attributable to Water, Sanitation, and Hygiene, by World Health Organization Subregions, 2002

| | | | | | Mortality stratum (thousands) | | | | |
| | | | | | Africa | | Americas | | |
	Number (thousands)	DALYs Attributable to WSH-Related Risk (%)[d]	Developed Countries (thousands)	Developing Countries (thousands)	High Child, High Adult	High Child, Very High Adult	Very Low Child, Very Low Adult	Low Child, Low Adult	High Child, High Adult
Population of children under 5 years	616,908		81,206	535,702	51,925	60,004	22,986	45,433	9,509
Total DALYs	433,832		13,004	420,828	82,044	85,893	2,299	15,358	5,063
Total WSH-related DALYs	107,881		786	107,095	26,568	26,178	39	1,397	990
Percentage of total DALYs	25		6	25	32	30	1.7	9.1	19.5
Diarrheal diseases[a]	47,520	44.0	432	47,088	9,283	9,390	22	901	565
Intestinal nematode infections[b]	531	0.5	1	530	117	58	0	15	38
Malnutrition (only protein-energy malnutrition)	7,104	6.6	70	7,034	1,239	1,412	0	244	104
Consequences of malnutrition[a]	28,475	26.4	166	28,309	6,853	5,744	0	95	177
Trachoma[b]	4	0.0	0	4	0	1	0	1	0
Schistosomiasis[b]	14	0.0	0	14	5	6	0	0	0
Lymphatic filariasis	146	0.1	1	146	17	18	0	0	0
Subtotal, WSH	83,794	77.7	669	83,125	17,516	16,629	22	1,256	884
Malaria	17,482	16.2	2	17,480	7,736	8,011	0	24	10
Onchocerciasis	5	0.0	0	5	3	2	0	0	0
Dengue	137	0.1	0	137	0	1	0	6	7
Japanese encephalitis	247	0.2	0	247	0	0	0	0	0
Subtotal, water resources management	17,871	16.6	2	17,868	7,739	8,014	0	30	17
Drownings	1,473	1.4	62	1,411	141	125	13	53	13
Subtotal, safety of water environments	1,473	1.4	62	1,411	141	125	13	53	13
Other infectious diseases[c]	4,743	4.4	52	4,690	1,172	1,411	5	59	76

Source: Data based on literature review/expert survey. Not a formal WHO estimate but based on Prüss-Üstün and Corvalán 2006.
Notes: DALY = disability-affected life year; WSH = water, sanitation, and hygiene. Subtotals may not add because of rounding.
a. Data further validated by Comparative Risk Assessment methods.
b. Included in the Comparative Risk Assessment study.
c. Not attributable to one group alone.
d. Percentage of all DALYs attributable to WSH-related risks.

Mortality stratum (thousands)								
Eastern Mediterranean		Europe			Southeast Asia		Western Pacific	
Low Child, Low Adult	High Child, High Adult	Very Low Child, Very Low Adult	Low Child, Low Adult	Low Child, High Adult	Low Child, Low Adult	High Child, High Adult	Very low Child, Very Low Adult	Low Child, Low Adult
15,213	53,234	21,974	17,576	10,859	28,449	150,232	7,812	121,704
5,086	53,622	1,719	5,717	2,670	10,738	117,012	599	45,507
614	12,911	37	600	88	2,038	28,390	22	8,010
12.1	24	2	10.5	3.3	19	24	4	18
412	6,500	21	342	39	1,120	14,896	8	4,020
0	40	0	0	0	21	108	0	134
81	834	0	48	22	316	1,938	0	853
52	3,482	0	163	3	333	9,774	0	1,784
0	1	0	0	0	0	0	0	0
0	2	0	0	0	0	0	0	0
0	2	0	1	0	5	96	0	9
546	10,861	21	554	64	1,794	26,810	9	6,800
18	665	0	2	0	158	715	0	142
0	0	0	0	0	0	0	0	0
2	3	0	0	0	17	53	0	48
0	50	0	0	0	7	51	0	139
20	719	0	2	0	182	819	0	329
24	123	5	24	17	35	246	4	652
24	123	5	24	17	35	246	4	652
24	1,208	11	21	7	26	514	9	229

Computing Country-Level Environmental Health Burden of Disease

THIS APPENDIX PROVIDES A DESCRIPTION OF THE METHODS used to estimate the environmental health burden of disease, including effects on education, and relative monetary costs.[1] An overview of the results for Ghana and Pakistan has been provided in chapter 5. The methodology (summarized graphically in figure D.1) has two main components. The first involves the valuation of mortality for each cause of death (left side of figure D.1), and the second relates to the estimation of educational costs of disease (right side of figure D.1).

Mortality

This section details the steps used to estimate mortality from environmental health risks (figure D.1, steps 1 to 6). The environmentally attributable disease in Ghana is estimated in reference to diarrhea, acute lower respiratory infections (ALRI), malaria, measles, protein-energy malnutrition, and a residual group of other causes of mortality (table D.1).

It is important to note that diarrhea constitutes both a cause of death and a risk factor through its effect on malnutrition and consequent mortality. Deaths related to urban air pollution are not included because of the small effects on child mortality that can be calculated from current estimation methods.

FIGURE D.1
Summary of the Methodology

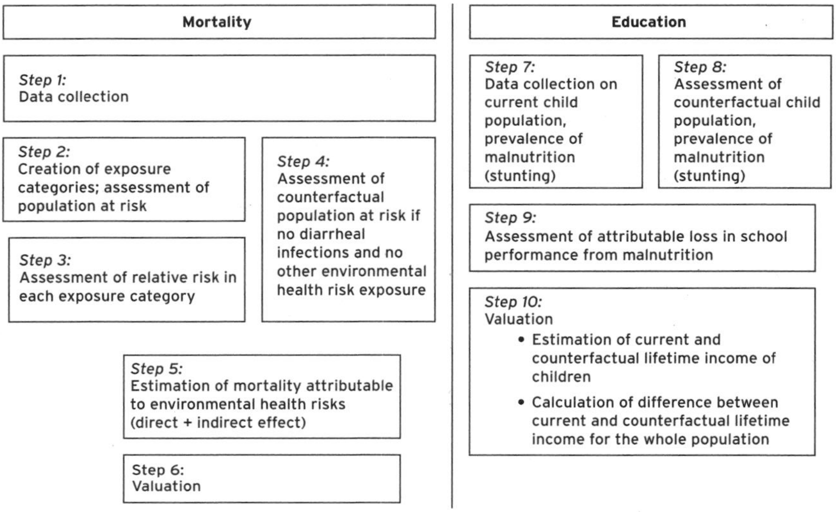

Mortality

Step 1: Data collection

Step 2: Creation of exposure categories; assessment of population at risk	Step 4: Assessment of counterfactual population at risk if no diarrheal infections and no other environmental health risk exposure
Step 3: Assessment of relative risk in each exposure category	

Step 5: Estimation of mortality attributable to environmental health risks (direct + indirect effect)

Step 6: Valuation

Education

Step 7: Data collection on current child population, prevalence of malnutrition (stunting)	Step 8: Assessment of counterfactual child population, prevalence of malnutrition (stunting)

Step 9: Assessment of attributable loss in school performance from malnutrition

Step 10: Valuation • Estimation of current and counterfactual lifetime income of children • Calculation of difference between current and counterfactual lifetime income for the whole population

Source: Compiled by World Bank team.

TABLE D.1
Causes of Death and Risk Factors Considered in this Study

Cause of Death	Direct Environmental Risk Factor	Indirect Risk Factor
Diarrhea	Inadequate water, sanitation, and hygiene	Malnutrition through diarrheal infections
Acute lower respiratory infection	Indoor air pollution	Malnutrition through diarrheal infections
Malaria	Poor management of water resources	Malnutrition through diarrheal infections
Measles	None[a]	Malnutrition through diarrheal infections
Protein-energy malnutrition	None[a]	Malnutrition through diarrheal infections
Other causes of mortality (excluding perinatal conditions)	None[a]	Malnutrition through diarrheal infections

Source: Compiled by World Bank team.
a. This report focuses on water, sanitation, and hygiene; indoor air pollution; and water resource management. None of these environmental health risks are considered to contribute significantly to these causes of death.

Total observed mortality—denoted as M^0—in any given population is the result of exposure to different risk factors. If the fraction of mortality caused by environmental risk factors is denoted as F^{EH}, then mortality attributable to environmental health risks, M^{EH}, is given by the formula

$$M^{EH} = \sum_{j=1}^{j=m} F_j^{EH} M_j^0, \tag{D.1}$$

where M_j^0 is current annual cases of cause-specific mortality; j is ALRI, diarrhea, malaria, measles, and "other causes" of deaths; F_j^{EH} is the fraction of cause-specific deaths j attributable to environmental health risks; and M^{EH} is the annual cases of mortality caused by environmental health risk factors.

The methodology for estimating M^{EH} relies on the estimation of the impact fraction F_j^{EH} for each cause of death j. A general formula for the fraction of deaths attributable to a set of risk factors is

$$F = \frac{\sum_{i=1}^{n} P_i RR_i - \sum_{i=1}^{n} P'_i RR_i}{\sum_{i=1}^{n} P_i RR_i}, \tag{D.2}$$

where the subscript i denotes exposure category, P_i is the population's current exposure distribution, P'_i is the population's counterfactual exposure distribution (that is, exposure distribution in the absence of environmental risks), and RR_i is the risk of mortality in each exposure category.[2]

Step 1: Data Collection

The estimation of the environmental health burden of disease and its subsequent monetary valuation uses data from a variety of sources. Typical sources of information are listed in table D.2. The more data are available from country sources, the better. As an alternative, one can use information from international sources, making the necessary assumptions to adapt the parameters to the country in question.

Step 2: Assessment of the Population at Risk

To compute the health burden of each cause of death or disease from environmental risk factors, one needs estimates of the population exposed to the relevant risk factors. Multiple environmental risk factors may contribute to a particular disease and mortality from this disease, and an environmental risk factor is often associated with several levels of risk. It is therefore desirable to create exposure categories, allowing for as many combinations of risk factors and levels of risk as is practically

TABLE D.2
Estimating the Cost of Environmental Health Risks: Information Types and Sources

Type of Information	Source	Source Level
Exposure to risk factors (for example, access to water, malnutrition, indoor air pollution)	Demographic and Health Surveys (DHSs), Multiple Indicator Cluster Surveys (MICSs), national nutrition surveys, national censuses, Living Standards Measurement Surveys (LSMSs)	Country-specific sources
Baseline health end points (for example, crude mortality rate, disease-specific mortality, disease prevalence)	Vital records, health information systems, and statistical estimates of cause-specific mortality; DHSs and MICSs; WHO national estimates of cause-specific mortality	Country-specific sources
Relative risks of illness and mortality: epidemiology literature	Example: the relative risk of dying from different levels of malnutrition is summarized in Fishman and others (2004)	Typically international studies and meta-analyses
Links between malnutrition and infections: epidemiology literature	Example: the effect of diarrheal diseases on weight gain in children is analyzed in Checkley and others (1997) and Molbak and others (1997)	International studies and meta-analyses
Value of mortality and morbidity	For the human capital approach: national accounts statistics (for example, gross domestic product), economic indicators (for example, wage rate) For the willingness-to-pay approach: value of a statistical life studies (benefit transfer approach)	For the human capital approach: country-specific sources For the willingness-to-pay approach: international studies and meta-analyses
Cost of illness	Economic indicators, public health statistics, interviews with hospitals and practitioners; DHSs and MICSs (treatment rates and type of treatment)	Country-specific sources

Source: Compiled by World Bank team.

feasible, and to compute the population shares that fall in each category. Figure D.2 presents a simplified illustration for ALRI mortality from indoor air pollution and malnutrition (see table D.1). In the case of diarrheal mortality, relevant risk factors include (a) lack of access to water, sanitation, or both and (b) malnutrition.

Exposure data can often be obtained from the Demographic and Health Surveys (DHSs). Ghana's 2003 DHS reports malnutrition levels for each child, and this information can be overlaid with household characteristics to obtain exposure to indoor air pollution and lack of water and sanitation.

FIGURE D.2
Exposure Categories

	Indoor air pollution	**No indoor air pollution**
Malnutrition	Exposure category 1	Exposure category 2
No malnutrition	Exposure category 3	Exposure category 4

Source: Compiled by World Bank team.

Step 3: Assessment of Relative Risks

Relative risks express the probability of an event (for example, developing a disease or dying from a disease) occurring in the exposed group versus the probability of the same event occurring in the group that was not exposed. Relative risks typically are available from the epidemiology literature. However, current research literature does not contain estimates of the combined relative risk of multiple environmental risk factors, such as the risk of death from exposure to indoor air pollution and from malnourishment, with the latter partially caused by diarrheal infections from inadequate water, sanitation, and hygiene. In this analysis, the problem is overcome by assuming a Cox proportional hazard model in the calculation of relative risks.[3] If that model is used, the relative risk is calculated as

$$RR_i = \exp(\gamma_1 ENV + \gamma_2 MAL),\qquad\qquad (D.3)$$

where RR is relative risk; ENV is a binary variable (that is, one that takes a value of either 1 or 0) indicating the presence of an environmental risk; MAL is a binary variable indicating the presence of malnutrition; and γ_1 and γ_2 are the parameters linking the presence of risk factors to the relative risk of a disease. It is easy to see that equation D.3 implies that the relative risk for exposure category 1 in figure D.2 is equal to the product of the relative risks for exposure categories 2 and 3 (that is, $RR_1 = RR_2 * RR_3$). The relative risks for exposure categories 2 and 3 are revealed in the epidemiology literature, which has been adjusted for potential biases.[4] Table D.3 summarizes.[5]

The relative risks discussed in literature could, overestimate the mortality from environmental risk factors if the effects of underweight on mortality in Fishman and others (2004) did not control for environmental factors or if the relative risks of mortality from environmental factors did not control for underweight status. An alternative approach used to obtain the results in chapter 5 is therefore applied to assess the potential magnitude of overestimation.[6] The approach uses relationships between least-squares coefficients when a variable is omitted from an equation. In the case of Ghana, the following two equations are estimated by using the 2003 DHS:

$$MAL = b_0 + b_1 ENV \qquad\qquad (D.4)$$

$$ENV = a_0 + a_1 MAL \qquad\qquad (D.5)$$

TABLE D.3
Relative Risks by Exposure Categories, Assuming Cox Hazard Model

	Environmental Health Exposure	No Environmental Health Exposure
Malnourished	$RR_1 = RR_2 * RR_3$	RR_2 from literature, adjusted (for example, Fishman and others 2004)
Not malnourished	RR_3 from literature, adjusted	$RR_4 = 1$

Source: Compiled by World Bank team.

The estimated coefficients b_1 and a_1 are then used in the following equations to solve for γ_i, where β_i represents the biased estimators in equation D.3:

$$\beta_2 = \gamma_2 + \gamma_1 * a_1 \tag{D.6}$$

$$\beta_1 = \gamma_1 + \gamma_2 * b_1 \tag{D.7}$$

Step 4: Calculation of Counterfactual Exposure Categories

The next step is to estimate the population in each exposure category in the absence of environmental risk factors. This step is important because child malnutrition depends not only on environmental health risks, but also on various nutritional factors such as maternal malnutrition and low protein-energy intake. Therefore, one needs to separate out the effects of nonenvironmental factors on malnutrition. Because the main environmental health risk influencing malnutrition acts through diarrheal infections, this analysis estimates malnutrition prevalence in a counterfactual child population that has not suffered from repeated diarrheal infections.

Estimation of the effect of diarrheal infections on nutritional status and consequent disease burden involves the following steps:

1. A weight gain retardation factor, β_t^i, is specified for each child i at age t and is a function of diarrheal infection dose from initial age.
2. The counterfactual weight, $W_t^{N_i}$ of child i at age t is estimated (that is, the weight of the child if the child had not suffered weight gain retardation from repeated diarrheal infections).
3. The weight-for-age z-score (WAZ) for $W_t^{N_i}$ and corresponding underweight malnutrition rates are calculated.
4. Mortality from environmental factors is then estimated.

The relationship between a child's observed weight, the weight gain retardation factor, and the counterfactual weight is

$$W_t^{D_i} - W_0^i = \left(1 - \beta_t^i\right) \times \left(W_t^{N_i} - W_0^i\right), \tag{D.8}$$

which gives

$$W_t^{N_i} = \frac{1}{\left(1-\beta_t^i\right)} W_t^{D_i} - \frac{\beta_t^i}{\left(1-\beta_t^i\right)} W_0^i,$$ (D.9)

where $W_t^{D_i}$ is the observed weight of child i at age t who has suffered weight gain retardation from diarrheal infections, and W_0^i is the initial weight of the child.

Table D.4 presents the weight gain retardation factors (for a subset of age groups) used in this study.[7] This factor is the reduction in weight gain from repeated infections over a period of a child's life in his or her first five years. For instance, if a child's initial weight was 5 kilograms at the age of 3 months and was 10 kilograms at the age of 40 months, and if the child had suffered a weight gain retardation of 25 percent ($\beta_t^i = 0.25$) during that period, then the child's weight would have been 11.67 kilograms in the absence of the weight gain retardation. In contrast, the median child at 40 months of age in the international reference population has a weight of 14.6 kilograms. The share of the weight retardation from repeated infections is therefore 36 percent.[8] The remaining weight retardation (64 percent) is from other factors.

The factors in table D.4 are used to calculate $W_t^{N_i}$ for each child in the DHS data set in Ghana, along with a new weight-for-age z-score for each child. The results imply that weight gain retardation from diarrheal infections is about 35 percent of total weight retardation relative to the median child in the international reference population (WAZ = 0) (see appendix A). This result is equivalent to approximately a 40 percent change in the standard deviation. This change in standard deviation is applied to each of the children in the Ghana DHS (for example, the new z-score is –1.8 for an old z-score of –3.0).[9]

TABLE D.4
Weight Gain Retardation Factors by Age and z-Score

Age (Months)	WAZ = −1	WAZ = −2	WAZ = −3	WAZ = −4
12	0.15	0.37	0.45	0.55
18	0.11	0.27	0.34	0.43
24	0.09	0.22	0.29	0.37
30	0.09	0.19	0.26	0.34
36	0.08	0.18	0.25	0.33
42	0.08	0.18	0.24	0.32
48	0.08	0.17	0.24	0.32
54	0.08	0.17	0.23	0.31
60	0.08	0.17	0.23	0.31

Source: Compiled by World Bank team.
Note: WAZ = weight-for-age z-score.

TABLE D.5
Weight for Age in Children under Five: Current Rates and Estimated Rates in the Absence of Diarrheal Infections in Ghana

Weight for Age Score	Prevalence Rates in 2003 (%)	Prevalence Rates in the Absence of Weight Gain Retardation from Diarrheal Infections (%)
> −1 WAZ (not underweight)	45.3	67.70
−1 to −2 WAZ (mild underweight)	32.9	30.00
−2 to −3 WAZ (moderate underweight)	17.3	2.30
< −3 WAZ (severe underweight)	4.6	0.04

Source: Compiled by World Bank team.
Note: WAZ = weight-for-age z-score.

Table D.5 presents current underweight prevalence rates in children under five from the 2003 DHS and estimated rates in the absence of weight gain retardation from diarrheal infections. In the absence of retardation from diarrheal infections, the estimates indicate that there would be practically no severely underweight children in Ghana. The prevalence of moderate underweight would also be substantially reduced, while mild underweight would be about the same as reported in the 2003 DHS.

Step 5: Estimation of Deaths Attributable to Environmental Health Risk Factors

The fraction of cause-specific deaths attributable to environmental risk factors (for example, the fraction of ALRI mortality attributable to indoor air pollution and diarrhea-related malnutrition) are obtained by combining in equation D.2 the parameters obtained in steps 2 and 3. This procedure is undertaken for each cause of death for which environmental risks are a contributing factor. Figure D.3 explains, using ALRI as an example. The left side of the figure presents child population shares and relative risks for the current situation, and the right side presents the counterfactual situation (that is, no environmental health risks). Combining this information in equation D.2 gives the impact fraction at the bottom of the figure.

Table D.6 presents the cause-specific mortality, values of impact fraction *F*, and mortality attributable to environmental risk factors for each cause of death in Ghana.

Step 6: Valuation of Attributable Health End Points

The last step consists of multiplying each end point by its unit cost. Mortality can be valued by first estimating a stream of income that the individual would have generated if he or she had been alive and then discounting the stream to obtain

FIGURE D.3
Exposure Categories, Population Shares, and Relative Risks of ALRI in Ghana

Current Situation			Situation with no Environmental Health Risks		
Population	Exposed to environmental health risk	Not exposed to environmental health risk	Population	Exposed to environmental health risk	Not exposed to environmental health risk
Malnourished	53.0%	1.7%	Malnourished	0.0%	32.3%
Not malnourished	41.6%	3.7%	Not malnourished	0.0%	67.7%

Relative risks	Exposed to environmental health risk	Not exposed to environmental health risk	Relative risks	Exposed to environmental health risk	Not exposed to environmental health risk
Malnourished	5.30	3.07	Yes	n.a.	2.10
Not malnourished	1.72	1.00	No	n.a.	1.00

Impact fraction
0.625

Source: Compiled by World Bank team.

TABLE D.6
Estimated Mortality in Children under Five from Environmental Risk Factors, Ghana

Cause of Mortality	Mortality (2005)	Impact Fraction (%)	Mortality Attributable to Environmental Health Risks
Pneumonia-ALRI	11,800	62.5	7,375
Diarrhea	9,900	91.0	9,009
Malaria	22,600	62.1	14,035
Measles	200	28.8	58
Protein-energy malnutrition	1,000	40.0	400
Other malnutrition-related causes of mortality (excluding perinatal conditions)	12,600	38.3	4,826
Total	58,100		35,703

Source: Compiled by World Bank team.
Note: Not including perinatal conditions. ALRI = acute lower respiratory infection.

a measure of lost productivity. This method—the *human capital approach*—is often considered to provide a lower-bound estimate of the value of avoiding mortality (see appendix F). Applying the human capital approach to valuation of mortality suggests a cost in Ghana equivalent to 5.55 percent of gross domestic product (GDP) for the effect of environmental factors.[10]

Education

The health links between malnutrition and the environment provide further insight into the issue of costing environmental health risks.[11] Repeated infections in early childhood contribute to child malnutrition, which, in turn, impairs cognitive development and educational performance. This effect has an impact on labor productivity, reducing individual earnings. Malnutrition can also lead to delayed primary school enrollment and grade repetition, which, in turn, may have an impact on lifetime income in terms of delayed labor-force entry.

To value the impacts of malnutrition through education, one needs to estimate the loss in education performance attributable to malnutrition, estimate the relationship between education and lifetime income, and combine this information to obtain the loss in lifetime income attributable to malnutrition. The discussion here begins with the data collection issues.

Step 7: Data Collection

Stunting, an indicator of chronic malnutrition, is often used when measuring the effects of malnutrition on education. Stunting is expressed as deviations from the median height for age z-score (HAZ) in an international reference population. Mild stunting means an HAZ between -1 and -2 standard deviations from the median child in the reference population ($-2 <$ HAZ < -1). Moderate stunting is defined as $-3 <$ HAZ < -2 and severe stunting as HAZ < -3. The current stunting prevalence rate in Ghana is presented in table D.7, row 1.

Step 8: Assessment of Counterfactual Malnutrition-Attributable Loss in School Performance

To assess the loss in school performance attributable to malnutrition, one first needs to establish the counterfactual situation. Table D.7 provides the counterfactual prevalence rates achievable by controlling for diarrheal infections. The share of severely malnourished children would go down from 10.5 percent (current situation) to 2.9 percent (situation without infections). Changes in prevalence rates are estimated by assuming a 25 percent change in standard deviation (consistent with the literature analyzed in appendix B and somewhat more conservative than the change used for underweight rates in table D.5).

The 25 percent change in standard deviation is arrived at in a way similar to that described in step 4. The effect of nonenvironmental factors on malnutrition first needs to be separated—through computations of the malnutrition prevalence of a counterfactual child population that has not suffered from repeated diarrheal infections. The estimation of the effect of diarrheal infections on stunting involves the actions described in step 4. A smaller retardation factor is used for height for age than for weight for age because it appears from the research literature

TABLE D.7
Estimated Annual Cost of Education Outcomes from Stunting and Share from Environmental Factors in Ghana

	Standard Deviations			
	> -1	-1 to -2	-3 to -2	< -3
Nutritional status (prevalence rate)				
1. Percentage of children (<5 years) stunted (height for age -1 standard deviation), current situation	42.4	28.20	19.00	10.50
2. Percentage of children (<5 years) stunted (height for age <-1 standard deviation), counterfactual situation	51.90	32.30	12.90	2.90
3. Standard deviation of median child in each nutrition bracket	n.a.	-1.50	-2.40	-3.50
Effects of malnutrition on education outcomes (years per malnourished child)				
4. Loss in grade attainment	0	0.50	1.00	1.50
5. Loss in learning productivity (grade equivalents)	0	0.80	1.60	3.60
6. Delayed labor-force participation [a]	0	0.42	0.85	1.29
Present value of lifetime income (US$ thousands per child) [b]				
7. Income at current malnutrition rates	42.3	21.6	12.6	6.5
8. Counterfactual income	42.3	23.0	13.8	8.4
Present value of lifetime income for birth cohort (US$ million) [b]				
9. Labor-force participation (% of population age 15-65)	75	75	75	75
10. Present value of lifetime income for birth cohort (US$ million), current malnutrition rates	6,985	2,999	1,207	373
11. Present value of lifetime income for birth cohort (US$ million), counterfactual income	6,985	3,183	1,325	479
12. Present value of lifetime income for birth cohort (US$ million), current malnutrition rates	11,564			
13. Present value of lifetime income for birth cohort (US$ million), counterfactual income	11,972			
14. Total cost of environmental factors (US$ million)	407			
15. Total cost of environmental factors (% of gross domestic product in 2005)	3.79			

Source: Compiled by World Bank team.
Note: n.a. = not applicable.
a. From delayed primary school enrollment and grade repetition.
b. Present value of lifetime income losses of an annual national birth cohort, adjusted for child mortality (discounted to childhood age = 1 year).

(see appendix B) that height growth retardation from infections is somewhat less than weight gain retardation (table D.8).

Using the factors in table D.8, one can calculate the counterfactual height (in the absence of growth retardation from infections) for each child in the DHS data sets in Ghana, along with a new height-for-age z-score for each child. The results imply that height growth retardation from diarrheal infections is 25 percent of total height retardation relative to the median child in the international reference population (HAZ = 0) (see appendix A). The new z-scores are 25 percent higher than observed z-scores (for example, the new z-score is −2.25 for an old z-score of −3.0).

Table D.9 presents current height-for-age prevalence rates in children under five from the 2003 DHS and estimated rates in the absence of height growth retardation from diarrheal infections. In the absence of height retardation from diarrheal infections, the estimates indicate a more than 70 percent reduction in the number

TABLE D.8
Height Growth Retardation Factors by Age and z-Score

Age (Months)	HAZ = −1	HAZ = −2	HAZ = −3	HAZ = −4
12	0.05	0.13	0.19	0.24
18	0.04	0.09	0.13	0.18
24	0.03	0.07	0.11	0.15
30	0.03	0.07	0.10	0.14
36	0.03	0.06	0.10	0.13
42	0.03	0.06	0.09	0.13
48	0.03	0.06	0.09	0.12
54	0.03	0.06	0.09	0.12
60	0.03	0.06	0.09	0.12

Source: Compiled by World Bank team.
Note: HAZ = height-for-age z-score.

TABLE D.9
Height-for-Age Rates in Children under Five: Current Rates and Estimated Rates in the Absence of Diarrheal Infections in Ghana

Height-for-Age (Malnutrition)	Prevalence Rates in 2003 (%)	Prevalence Rates in the Absence of Height Growth Retardation from Diarrheal Infections (%)
> −1 standard deviation (not stunted)	42.4	51.9
−1 to −2 standard deviation (mild stunting)	28.2	32.3
−2 to −3 standard deviation (moderate stunting)	19.0	12.9
< −3 standard deviation (severe stunting)	10.5	2.9

Source: Compiled by World Bank team.

of severely stunted children and a more than 30 percent reduction in the number of moderately stunted children. The prevalence of mild stunting would be about the same as reported in the 2003 DHS.

Step 9: Assessment of Malnutrition-Attributable Loss in School Performance

A number of recent longitudinal studies (see appendix B) have investigated the effect of early childhood malnutrition on educational outcomes. These studies use data that follow children over time, usually from birth well into primary or secondary schooling, containing various anthropometric, schooling, and socioeconomic status indicators. The educational and cognitive impacts of malnutrition are expressed as changes in

- Final grade attainment
- Learning productivity or school achievements (often measured by cognitive and math and reading achievement tests)
- School grade repetition
- Years of delayed primary school entry

The studies reviewed show that the change in grade attainment ranges between 0.12 years (mild stunting in the Philippines) and 0.91 years (moderate malnutrition in Brazil) and 0.7 years (for each z-score in Zimbabwe). The effect of malnutrition on learning productivity or achievement appears to be even larger, ranging from 0.8 years of grade equivalent in Guatemala and the Philippines for each z-score to 1.8 years for each z-score in severely malnourished children in those countries. Delayed primary school entry varies between two to four months per one z-score in the studies reviewed. The parameters used for Ghana are summarized in table D.7.

The magnitude of the effect of a child's nutritional status on grade attainment applied to Ghana (row 4, table D.7) is somewhat lower than that in the study of Zimbabwe, about the same as that in the study of Brazil, but significantly higher than that in the study in the Philippines. The rationale for applying a magnitude higher than that found in the Philippines is the substantially higher primary school completion rate there. One important reason for that higher rate is the higher emphasis on school completion in the Philippines, which likely reduces losses in grade attainment of children with lower nutritional status.[12] Primary school competition and years of children's school life expectancy in Zimbabwe are also higher than in Ghana. Choosing a lower effect on grade attainment in Ghana than that found in Zimbabwe may therefore be considered conservative and possibly an underestimate of actual effect.

The magnitude of effect of nutritional status on school achievement or learning productivity applied to Ghana (row 5, table D.7), measured in grade equivalents, is about the same as found in Guatemala and the Philippines for children with mild and moderate height-for-age deficit or stunting. For children with severe stunting, the magnitude applied is about the midpoint of that found in the Philippines and Guatemala.

Delayed primary school enrollment is derived from the study in Zimbabwe. Grade repetition is half of that found by Daniels and Adair (2004) in the Philippines. The combined effect of delayed enrollment and grade repetition is reported in row 6, table D.7.

The cost of delayed enrollment and grade repetition in terms of delayed labor-force participation is, however, very minimal compared with the cost of lower grade attainment and learning productivity. So variations in the magnitude of effects, within the range found in studies, have minimal influence on the estimated total cost of nutritional status.

Step 10: Valuation

To arrive at costs related to educational and cognitive impacts, one has to value the impact of lower grade attainment, learning productivity, and delay in primary school entry on lifetime income. To do this, one has to

1. Estimate private marginal returns to schooling—that is, a measure of the increment in returns that one extra year of schooling allows. This measure helps to estimate the lost returns for one less year of schooling (owing to lower educational attainment, for example).
2. Estimate the social cost of providing a year of schooling.
3. Using these estimates, calculate the marginal increase in income that schooling allows.
4. Estimate the lifetime income in the current situation and in the absence of infections (shown in rows 7 and 8, table D.7).

The present value of lifetime income of a child in the current situation is

$$PV(I) = \sum_{t=15}^{t=65} I_0 \left[(1+g)/(1+r)\right]^t, \qquad (D.10)$$

where I_0 is initial annual income of adults at the time the child is one year old, g is annual rate of growth in real per capita income, and r is a constant annual discount rate. It is assumed that an individual has a productive working life starting at the age of 15 ($t = 15$) and ending at the age of 65 ($t = 65$) (see the Ghana DHS).

The present value of counterfactual lifetime income of a child in the absence of infections is

$$PV(I^C) = PV(I) \left[1 + \delta Z \sum_{j=1}^{j=3} i_j \, \delta E_j\right], \qquad (D.11)$$

where δZ is the counterfactual change in height-for-age z-score; i_j is the percentage change in present value of lifetime income from an additional year of schooling

($j = 1$), improved learning productivity equivalent to a year of schooling ($j = 2$), and an additional year of labor force participation ($j = 3$); and δE_j is the change in education outcome and labor-force participation, for $j = 1, \ldots, 3$, per one z-score.

Specifically, δZ is the absolute value of 25 percent of the current z-score of each child. The percentage change in present value of lifetime income from an additional year of schooling and from improved learning productivity equivalent to a year of schooling is estimated from table D.10. For change in lifetime income from an additional year of school ($j = 1$)—that is, increased grade attainment—the cost of this additional year of schooling (C) is subtracted from income at age 15.[13] The percentage change in lifetime income from the change in entry into the labor force is estimated on the basis of lifetime income flows.

Change in educational outcome and labor-force participation for one z-score is as follows (see table D.7):

- 0.5 years of grade attainment (row 4, table D.7)
- 0.8 years of grade equivalent from learning productivity (school achievement) for children with mild and moderate stunting (row 5, table D.7)

TABLE D.10
Parameter Values Applied in Estimation of Income Losses

Parameters	Ghana	Pakistan	Remarks
Average annual income per income earner, 2005	₵4,370,000	PRs 57,800	Ghana: GDP per capita in 2005 (a proxy for per capita income); Pakistan: income in 2001/02 (Household Integrated Economic Survey 2001/02), adjusted to 2005
r	3%	3%	Same as used by WHO for health outcomes
g	2%	2%	Assumed per capita annual real income growth over the next 65 years
C_t	$0.37\,I_0(1+g)^{t-1}$	$0.37\,I_0(1+g)^{t-1}$	Assumption that cost of a year of schooling is 37% of per capita income at time $= t$
i_1	9.5%	7.2%	Calculated from rate of return to a year of schooling; that is, 11.7% in Sub-Saharan Africa and 9.9% in Asia (see Psacharopoulos and Patrinos 2004), less the cost of an additional year of schooling
i_2	13.2%	10.9%	Calculated from rate of return to a year of schooling; that is, 11.7% in Sub-Saharan Africa and 9.9% in Asia (Psacharopoulos and Patrinos 2004)

Source: Compiled by World Bank team.

- 2 years of grade equivalent from learning productivity (school achievement) for children with severe stunting, for the first one-unit improvement in z-score (row 5, table D.7)
- 0.42 to 0.44 years' change in entry into the labor force (row 6, table D.7)

Additional parameter values are presented in table D.10. Psacharopoulos and Patrinos (2004) reviewed studies of returns to investment in education. Returns to one additional year of schooling are given by the coefficient on years of schooling. On average, the return to one additional year of schooling is 9.7 percent. The return is 11 to 12 percent in Sub-Saharan Africa and the group of low-income countries. This figure is used to estimate the percentage change in the present value of lifetime income from improved educational performance in the absence of early childhood diarrheal infections.

Four educational outcomes are considered:

1. A one-year lower grade attainment caused by poor nutritional status (stunting) results in an additional year of income (leaving school early) at $t = 15$, loss of income in all years of working life (lower income per year) for each $t \geq 16$, and a saving in cost of a year of schooling at $t = 15$.[14] $C_t = 0$ for $t \neq 15$.
2. Lower learning productivity equivalent to one year of schooling results in a loss of income in all years of working life for each $t \geq 15$. $C_t = 0$ for all years.[15]
3. Delayed primary school enrollment results in delayed labor-force entry. A one-year delay results in lost income at $t = 15$. $C_t = 0$ for all years.[16]
4. Grade repetition has the same effect as delayed enrollment.[17]

The cost of these educational outcomes is implicit in the difference between current and counterfactual lifetime incomes in rows 7 and 8, table D.7.[18] In fact, counterfactual lifetime income in each malnutrition category rises as a consequence of reducing infections and subsequent higher educational outcomes.

In estimations of current and counterfactual lifetime income for children of different nutritional status (from not malnourished to severely malnourished), assumptions about baseline income (I_0 in equation D.10) are important. Malnourished children live in disproportionately poorer households, and poverty is correlated with lower socioeconomic status, such as lower educational attainment of parents. This factor, in turn, is correlated with lower educational attainment of children as they grow up. Malnourished children are therefore likely to have lower income than children who are not malnourished, not only because of the effect of malnutrition on educational performance and income but also because of the socioeconomic status of the household they grew up in.

Therefore, instead of using GDP per capita as a proxy for per capita income for all children, this analysis attempts to develop a matrix of income differentials in Ghana across the range of nutritional status, from not malnourished to severely malnourished. These figures represent a more accurate estimate of lifetime income of children as they move across nutritional categories. Table D.11, panel a, shows the distribution of children by nutritional status and household wealth in Ghana.

TABLE D.11
Income Distribution across Malnutrition Categories and Wealth Quintiles in Ghana

Wealth Quintile	Not Malnourished (Height-for-Age > −1)	Mild (Height-for-Age −1 to −2 Standard Deviation)	Moderate (Height-for-Age −2 to −3 Standard Deviation)	Severe (Height- for-Age < −3 Standard Deviation)	Total [b]
a. Height-for-age prevalence rates (Ghana 2003 DHS)					
Poorest	30.04	29.1	25.47	15.39	100
Poorer	37.68	30.85	19.12	12.36	100
Middle	40.77	29.27	20.45	9.51	100
Richer	49.49	28.12	15.84	6.55	100
Richest	65.24	20.96	8.9	4.9	100
Total	42.42	28.15	18.95	10.49	100
b. Per capita income (expenditure) shares (%)[a]					
Poorest	1.7	1.6	1.4	0.9	5.6
Poorer	3.8	3.1	1.9	1.2	10.1
Middle	6.1	4.4	3.0	1.4	14.9
Richer	11.3	6.4	3.6	1.5	22.9
Richest	30.4	9.8	4.1	2.3	46.6
Total	53.3	25.3	14.2	7.3	100.1
c. Per household income (expenditure) shares (%)[a]					
Poorest	2.8	2.7	2.4	1.4	9.4
Poorer	6.4	5.2	3.2	2.1	16.9
Middle	7.8	5.6	3.9	1.8	19.2
Richer	12.7	7.2	4.0	1.7	25.6
Richest	18.9	6.1	2.6	1.4	28.9
Total	48.6	26.8	16.2	8.5	100.0

Source: Compiled by World Bank team.
a. Estimated from the height-for-age prevalence rates and total income (expenditure) shares.
b. Total income (expenditure) shares by quintile in Ghana in 1998/99 from the Ghana Living Standard Survey (World Bank 2007h; Trades Union Congress 2004).

Panels b and c show the income (expenditure) shares in relation to household wealth quintile and children's nutritional status. Total income (expenditure) shares for each group of children (not malnourished and mild, moderate, and severely malnourished) are then used to estimate baseline annual income (I_0) for each of these four groups, which is applied in equations D.10 and D.11 to estimate lifetime income (rows 7 and 8, table D.7).

By combining the information on current stunting prevalence (row 1, table D.7), lifetime income (rows 7 and 8), and labor-force participation (row 9), one can obtain total current and counterfactual lifetime income per malnutrition category (rows 10 and 11) and for the whole population birth cohort (rows 12 and 13).

The difference between current and counterfactual lifetime income for an annual population birth cohort as a whole represents the educational and cognitive

TABLE D.12
Annual Cost of Environmental Factors (Percentage of GDP in 2005), Using 3 Percent Discount Rate

	Ghana			Pakistan		
	Annual Deaths	Cost (₵ million)	Cost (US$ million)	Cost (% of GDP in 2005)	Annual Deaths	Cost (PRs billion)	Cost (US$ million)	Cost (% of GDP)
Estimates excluding malnutrition-mediated effects								
Mortality effects	24,712	371	412	3.84	131,611	195	3,250	2.90
Estimates including malnutrition-mediated effects								
Mortality effects	35,702	537	595	5.55	187,429	278	4,633	4.13
Education effects		367	407	3.79		317	5,281	4.71
Total annual costs		904	1,002	9.34		595	9,914	8.84

Source: Compiled by World Bank team.
Note: ₵ = Ghanaian new cedi.

TABLE D.13
Annual Cost of Environmental Factors (Percentage of GDP in 2005), Using 5 Percent Discount Rate

	Ghana				Pakistan			
	Annual Deaths	Cost (¢ million)	Cost (US$ million)	Cost (% of GDP in 2005)	Annual Deaths	Cost (PRs billion)	Cost (US$ million)	Cost (% of GDP)
Estimates excluding malnutrition-mediated effects								
Mortality effects	24,712	188	208	1.94	131,611	99	1,650	1.46
Estimates including malnutrition-mediated effects								
Mortality effects	35,702	271	300	2.80	187,429	140	2,333	2.09
Education effects		180	199	1.86		156	2,592	2.31
Total annual cost		451	499	4.66		296	4,925	4.40

Source: Compiled by World Bank team.
Note: ¢ = Ghanaian new cedi.

cost of malnutrition from diarrheal infections. This cost is estimated at 3.8 percent of GDP in 2005. The cost attributable to environmental factors (that is, diarrheal infections from inadequate water supply, sanitation, and hygiene) is about one-third of the total cost of stunting on lifetime income.

Summary of Results for Ghana and Pakistan

The methodology described in this appendix has also been applied to Pakistan. The results, summarized in chapter 5, are reported here for convenience (table D.12.). In total, the cost of malnutrition-mediated effects (mortality and education effects) of environmental risk factors on children in Ghana and Pakistan is estimated at an equivalent of 9.2 percent and 8.6 percent, respectively, of GDP per year. This finding is in stark contrast to a cost of 3.8 percent and 2.9 percent, arrived at when considering only the estimates excluding malnutrition-mediated effects of environmental risks.

The annualized monetary cost of mortality and impaired education is sensitive to the discount rate of income losses. Annual costs are presented in tables D.12 and D.13 using discount rates of 3 and 5 percent. In both Ghana and Pakistan, at an annual discount rate of 5 percent, the annual cost is half of the cost at a 3 percent discount rate. Total costs in Ghana and Pakistan are, however, nearly two and one-half and nearly three times higher, respectively, than the cost that does not include malnutrition effects, regardless of the discount rate used.

Notes

1 This appendix is based on the analysis of country-specific data and surveys for Ghana. Results are arrived at using a methodology developed specifically for this report by Björn Larsen (consultant). Useful input was provided by Maureen Cropper. The analysis estimates costs that are attributable to environmental risks in children under five years of age for Ghana and Pakistan and for the first time includes the malnutrition-mediated effects, as well as the longer-term impacts on cognition, school performance, and future work productivity.

2 See appendix E for more detail on impact fractions or attributable fractions.

3 The use of the Cox hazard model has a drawback if one wants to calculate the attributable fractions of individual risks (as opposed to calculating joint attributable fractions, as is done here). In fact, in such model specifications, the effect of each risk factor is augmented by the multiplicative interaction with other risk factors. The result is that the fractions attributable to individual risks may add to more than 1. If one imposes additive relative risks (so that the sum of the attributable fraction equals 1), the results do not differ much from those presented here.

4 The estimate of the impact of a single risk factor on a disease (for example, the impact of malnutrition on deaths caused by ALRI) is often biased because other risk factors were not controlled for in the epidemiological study. So if the relative risk of ALRI death attributable to malnutrition is estimated ignoring indoor air pollution, MAL will pick up its effects, because malnutrition and exposure to indoor air pollution are correlated in most populations. Appendix E describes a method to deal with biased relative risks.

5 This framework for calculating relative risks can be expanded to a larger set of exposure categories that reflect different risk levels associated with *ENV* and *MAL*. For instance, indoor air pollution exposure categories can be households using fuelwood, charcoal, unimproved rather than improved stoves, and liquefied petroleum gas. Malnutrition can be mild, moderate, and severe.

6 For the alternative approaches, please see appendix E.

7 Although the research literature finds that the retardation factor often is in the range of 0.2 to 0.5, there is limited research on how the retardation factor may vary with z-score and age. It is postulated here that the factor declines with age and is greater the more underweight the children are. Diarrheal infections are most frequent during the ages of 6 to 11 months and 12 to 23 months. It may therefore be expected that weight gain retardation from diarrheal infections is greatest in this period and, at least in Ghana and Pakistan, is particularly great up to the age of 12 months, during which underweight increases rapidly. The retardation factors in table D.4 for WAZ = −2 to −4 are within the bounds often found in the research literature.

8 The calculation is as follows: $(11.67 − 10)/(14.6 − 10) = 36\%$.

9 The calculation is as follows: $−3 + (3 * 0.4) = −1.8$.

10 A working life from 15 to 65 years is applied. Annual income is approximated by GDP per capita. Income is discounted at 3 percent. Real income growth is 2 percent per year.

11 The methodology for this part of the analysis, which estimates effects of stunting (partly attributed to repeated infections) on cognition and learning, has benefited from comments by Harold Alderman (senior adviser, Africa Region Human Development Department, World Bank).

12 On-time enrollment, grade attainment, and high school completion rates in Cebu are relatively high: 94 percent on-time (less than 8 years) enrollment, 9.5 years of schooling, and 55 percent high school completion rate.

13 An average child in Ghana starts school at the age of about seven and has seven to eight years of schooling (World Bank 2007h). The average age of leaving school is therefore around 15 years.

14 Cost of schooling is the private and public costs of schooling.

15 Notice that there is a possibility of underestimation of damages linked to impaired cognitive development. In fact, income earners with impaired cognitive development are more likely to have lower income than others with the same educational achievement. The negative wage effect for income earners with impaired cognitive development as a consequence of early childhood malnutrition may therefore be larger than that reflected in the studies reviewed by Psacharopoulos and Patrinos (2004), if those studies did not control for cognitive functioning.

16 Delayed school enrollment may affect grade attainment and thus schooling cost. However, including such an effect here would involve double counting because grade attainment is dealt with separately.

17 Increased grade repetition as a consequence of poor nutrition status is equivalent to an average of one month of a school year per child with severe stunting status and one week per child with mild stunting status. The additional schooling cost of grade repetition is therefore ignored here because it is minimal compared with the costs related to grade attainment.

18 These amounts are estimated from the Ghana 2003 DHS.

Methodological Aspects of Assessing Environmental Health Burden of Disease

THIS APPENDIX CONSISTS OF TWO SECTIONS that provide details on certain methodological aspects of the analysis to estimate the environmental health burden of disease.[1] The first section relates to how one gets from relative risks (RRs) to attributable fractions (AFs). Environmental risk factor AFs for Ghana and Pakistan are discussed in chapter 5, and a detailed methodology is described in appendix D. The second section discusses a possible way to get around the biased estimates of RRs. Unbiased estimates of RRs are used to calculate the AFs in table 5.5.

From Relative Risks to Attributable Fractions

The burden of disease methodology aims at estimating the effects of selected risk factors on mortality and disease. It applies disability weights to combine deaths and illness into a single metric: disability-adjusted life years (DALYs). The environmental burden of disease is obtained by limiting the analysis to environmental risk factors.[2] The World Health Organization's Comparative Risk Assessment (WHO 2002) considers six environmental risk factors and puts them in relation to the diseases identified in the table E.1.

For each pair of risk factors and diseases, the Comparative Risk Assessment defines a relative risk—that is, the probability of a disease or death from that disease (for example, diarrhea) occurring in a group exposed to a risk factor

TABLE E.1
Environmental Risk Factors and Related Diseases Included in the WHO Comparative Risk Assessment

Risk Factor	Related Diseases
Outdoor air pollution	Respiratory infections, selected cardiopulmonary diseases, lung cancer
Indoor air pollution	Chronic obstructive pulmonary disease, lower respiratory infections, lung cancer
Lead	Mild mental retardation, cardiovascular diseases
Water, sanitation, and hygiene	Diarrheal diseases, trachoma, schistosomiasis, ascariasis, trichuriasis, hookworm disease
Climate change	Diarrheal diseases, malaria, selected unintentional injuries, protein-energy malnutrition
Selected occupational risk factors:	
Injuries	Unintentional injuries
Noise	Hearing loss
Carcinogens	Cancers
Airborne particulates	Asthma, chronic obstructive pulmonary disease
Ergonomic stressors	Lower back pain

Source: WHO 2002.

(for example, inadequate water, sanitation, and hygiene) versus the probability of the same event occurring in a group that was not exposed.

$$RR = \frac{Pr\left(Disease \mid Exposed\right)}{Pr\left(Disease \mid Nonexposed\right)} \qquad (E.1)$$

So, for example, a relative risk of 2.3 of mortality from ALRI from indoor air pollution (IAP) means that the probability of a person contracting ALRI after being exposed is 2.3 times higher than the probability of the same person contracting ALRI if not exposed. RRs, which are typically obtained from epidemiology studies and expert forums, are reported for both mortality and morbidity and constitute the basic building block of the estimation of burden of disease.

Information on exposure to a risk factor (for example, the number of children under five years of age exposed to IAP) is combined with the RRs to estimate the risk AF of death or illness. For example, suppose one wants to attribute deaths from acute respiratory infections (ARI) in children under five to exposure to IAP. The standard definition of the fraction of ARI deaths attributable to IAP is

$$AF = (\text{Excess deaths attributable to IAP})/(\text{Total ARI deaths}) \qquad (E.2)$$

Excess deaths attributable to IAP
= (Deaths in exposed population based on RR)
− (Deaths in exposed population if no exposure)   (E.3)

The fraction of deaths attributable to IAP is the ratio of excess deaths caused by that pollution to total ARI deaths. Excess deaths attributable to IAP is the difference between deaths in the exposed population based on estimated RR, minus deaths in the exposed population if there were no exposure—that is, if the RR were 1.

The AF defined in equations E.2 and E.3 can be written as follows. Let N equal the population of interest (for example, children under five in Ghana) and P be the proportion of the population exposed to the risk factor (for example, IAP). Let m be the death rate from ARI in the unexposed population; RR^*m is the death rate from ARI in the exposed population. Then, from equation E.3,

$$\text{Excess deaths attributable to IAP} = NP(RR^*m) - NPm$$
$$= NP(RR - 1)m \tag{E.4}$$

$$\text{Total deaths attributable to ARI} = NP(RR^*m) + N(1 - P)^*m$$
$$= NP(RR - 1)m + Nm$$

Dividing the right-hand side of equation E.4 by total deaths gives

$$\text{AF from IAP} = [NP(RR - 1)m]/\,[NP(RR - 1)m + Nm] \tag{E.5}$$

Canceling N and m from the numerator and denominator yields the standard formula:

$$\text{AF from IAP} = P(RR - 1)/[P(RR - 1) + 1] \tag{E.6}$$

Equation E.6 generalizes when there is more than one exposure category (for example, exposure to indoor smoke in an unventilated kitchen and exposure to indoor smoke in ventilated kitchen) to

$$AF = \frac{\displaystyle\sum_{i=1}^{n} P_i RR_i - 1}{\displaystyle\sum_{i=1}^{n} P_i RR_i}, \tag{E.7}$$

where P_i is the population exposed to level i of the risk factor and RR_i is the relative risk of dying from the selected disease by being exposed to level i of the risk factor. It is important to note that equation E.7 also applies when there are multiple risk factors and when exposure categories represent different combinations of exposure to risk factors.

A generalized version of equation E.7 is

$$AF = \frac{\displaystyle\sum_{i=1}^{n} P_i RR_i - \sum_{i=1}^{n} P_i' RR_i}{\displaystyle\sum_{i=1}^{n} P_i RR_i}, \tag{E.8}$$

where the subindex i denotes exposure category; P_i is the population's current exposure distribution; P_i' is the population's exposure distribution after controlling one or more risk factors (that is, exposure distribution in the absence of environmental risks); and RR_i is the risk of mortality in each exposure category. A major advantage of equation E.8 is that it allows estimation of the AF of a target risk factor (such as IAP) while leaving other risk factors (such as malnutrition that is not environmentally attributable) unchanged. This advantage is exploited in estimating the burden of disease for Ghana and Pakistan in this report.

Dealing with Biased Estimates of Relative Risk

There is a second problem relating to this study in estimating both the direct and indirect effects of environmental risk factors. The RRs of mortality from exposure to IAP and from malnutrition are usually calculated in separate epidemiological studies, in which the other risk factor is not controlled for. For example, most of the studies of ALRI that is caused by IAP that are reviewed in Desai, Mehta, and Smith (2004) do not control for nutritional status. In the study by Fishman and others (2004) of the impact of malnutrition on ALRI, there is no control for IAP. In econometric jargon, this problem is referred to as the *omitted variable bias problem*.

If there is positive correlation between malnutrition and exposure to IAP, the RR in each of the studies is likely to overstate the impact of the risk factor studied, relative to a study in which both risk factors are included. This problem is difficult to correct; a second-best approach is suggested here, which is based on the properties of least-squares estimators in the presence of omitted variables.

The estimates in Fishman and others (2004) of the effect of malnutrition on mortality from various diseases are very likely biased because they fail to control for other risk factors.[3] If the estimate of the relative risk of death from ALRI attributable to malnutrition ignores IAP, malnutrition will pick up its effects, because malnutrition and exposure to IAP are correlated in most populations. For policy purposes, there are at least four risk factors whose impact on ALRI deaths might need to be estimated: exposure to IAP,[4] malnutrition, crowding, and poor medical treatment received when ill.

These factors are all important, and all have policy significance. Because poor children are likely to be at risk for all four factors, omitting three of them makes it likely that malnutrition will pick up their effects, thus overstating the impact of malnutrition. In most studies, one could reduce this bias by using family income (or assets) or mother's education as a proxy for the omitted factors, even if the factors could not be estimated directly. The only way to solve this omitted variable bias problem is to find studies that have controlled for other risk factors (or have proxied for them) in addition to malnutrition.

To understand the nature of the bias, suppose that in the correct model

$$y = \gamma_0 + \gamma_1 IAP + \gamma_2 MAL, \quad (E.9)$$

where y = death rate attributable to ALRI or the logarithm of the death rate.

It is well known that if malnutrition (MAL) is dropped from this equation and

$$y = \beta_0 + \beta_1 IAP \quad (E.10)$$

is estimated instead, that the least-squares estimator of β_1 —that is, $\hat{\beta}_1$—is related to the least-squares estimators of the coefficients in equation E.9 as follows:

$$\hat{\beta}_1 = \hat{\gamma}_1 + \hat{\gamma}_2 \hat{\delta}, \quad (E.11)$$

where $\hat{\delta}$ is the coefficient of IAP in a regression of MAL on IAP.

This equation implies that if $\hat{\gamma}_2 > 0$ and $\hat{\delta} > 0$, $\hat{\beta}_1$ will be biased upward: indoor air pollution will capture the effects of malnutrition. This situation is likely in practice, since indoor air pollution and malnutrition are positively correlated in most populations.

A similar argument can be made for a model in which IAP is dropped from equation E.9. In such a case, the coefficient for MAL will be biased as it captures the effect of IAP. In the calculation of the total cost of environmental health risk in chapter 5, relative risks are corrected by using these econometric results. Although this solution is second best, it is done in order to provide conservative estimates.

Notes

1 This appendix is based on detailed comments and guidance from Maureen Cropper.

2 To distinguish environmental risk factors from nonenvironmental risk factors, one has to define *environment*. This report follows Prüss-Üstün and Corvalán (2006: 22) by defining environment as "all the physical, chemical, and biological factors external to a person, and all the related behaviors."

3 Pelletier (1994) has demonstrated that estimates are robust when a wide array of variables is controlled for.

4 Notice that IAP could also contribute to malnutrition through repeated respiratory infections. Multicollinearity will cause the estimated effect to be less robust (although unbiased).

Monetary Valuation of the Cost of Environmental Health Risks

CHAPTER 5 PROVIDES MONETARY ESTIMATES of the environmentally attributable burden of disease in Ghana and Pakistan. The World Health Organization's methodology for estimating burden of disease is a powerful tool for decision-makers. Knowing the burden of disease makes it possible to compare policy options that act on different causes of death and have different mortality and morbidity impacts. A major advantage of using burden of disease is that it requires no monetary valuation of the benefits of an intervention (for example, there is no need to know the value of a life).

The main limitation of burden of disease is that it cannot be applied when one compares policies with outcomes measured in different metrics (for example, a case in which one policy reduces diarrhea mortality while another policy increases children's cognition and, thus, future income). In such cases, the monetary valuation of burden of disease can be very useful. Monetary valuation has also been used extensively to express burden of disease in a metric that is easy for policy-makers with no background in public health to understand.

Several approaches are available to value the impact of health risks. Traditionally, valuation of health damages has been based on the calculation of the observed costs caused by death or illness.

- Death, incapacity, or illness is typically valued using the human capital approach, which calculates the present value of forgone earnings.[1] For the sake of simplicity, many studies of the costs of environmental degradation have valued a lost year of healthy life using gross domestic product per capita as a proxy of the forgone annual earnings.

- Illness also results in expenses other than lost wages. The cost-of-illness approach measures such costs by estimating the change in costs incurred as a result of a change in the incidence of a particular illness. Both direct costs (such as cost of doctor visits and treatment) and indirect costs (such as loss of wages) are included in the estimation. In cases where some of the costs are borne by medical insurance, cost-of-illness measures will not be limited to a patient's out-of-pocket expenses but should include the additional costs borne by the insurance company or treatment facility, to capture the social benefits of the reduced risk. It is important to keep in mind that unit values should be expressed net of taxes and subsidies. Taxes and subsidies are transfers within the economy and should not be taken as a component of the true economic value of an item. So, for example, admissions in public hospitals that are free (or subsidized) can be valued by using a comparable estimate from the private sector. Likewise, measures of wage are taken net of labor taxes.

More recently, economists have started adopting estimates of willingness to pay (WTP) or willingness to accept compensation to value health damages. The human capital and cost-of-illness approaches take into account only the out-of-pocket expenses paid as a consequence of illness or death. They are based on the notion that individuals should at least be willing to pay those out-of-pocket expenses to avoid illness. But individuals' willingness to pay to avoid illness or death may be higher than their out-of-pocket expenses. The WTP approach, when correctly applied, can capture people's preferences to avoid pain and discomfort. It does so by looking at how they respond to risk of illness or death. Typically, valuation is determined either by observing behavior or by eliciting responses from a questionnaire. Responses are then analyzed using econometric techniques to estimate willingness to pay.

Despite its advantages, the WTP approach produces a much higher uncertainty regarding values and requires much closer scrutiny. The WTP approach is also data intensive: proper estimates require large and costly surveys and careful application of econometric methods. For this reason, studies in developing countries have commonly transferred WTP estimates from U.S. and European studies by accounting for differences in purchasing power and, occasionally, the income elasticity of consumption.

An important limit to the application of the WTP approach is with regard to children's life. It is conceptually and practically very difficult to observe or measure monetarily children's preferences. For this reason, in this study, valuation relies solely on the human capital approach. For simplicity, the study does not look at cost of illness and it concentrates on the valuation of mortality. Deaths in Ghana are valued using a per capita income of US$485 (2005 dollars). In Pakistan, a per capita income of US$714 (2005 dollars) is used. Per capita income values are also used to estimate the present value of income losses from lower education attainment, together with an assumed real rate of income growth of 2 percent.

Note

1   Some have argued that there is a conceptual problem with the human capital approach, in that most people value safety not out of concern for preserving current and future income levels but primarily because they have an aversion to pain, suffering, and death.

References

Adair, L., B. Popkin, J. VanDerslice, J. Akin, D. Guilkey, R. Black, J. Briscoe, and W. Flieger. 1993. "Growth Dynamics during the First Two Years of Life: A Prospective Study in the Philippines." *European Journal of Clinical Nutrition* 47 (1): 42–51.

Adjuik, M., T. Smith, S. Clark, J. Todd, A. Garrib, Y. Kinfu, K. Kahn, M. Mola, A. Ashraf, H. Masanja, U. Adazu, J. Sacarlal, N. Alam, A. Marra, A. Gbangou, E. Mwageni, and F. Binka. 2006. "Cause-Specific Mortality Rates in Sub-Saharan Africa and Bangladesh." *Bulletin of the World Health Organization* 84 (3): 181–88.

Agarwal, B. 1983. "Diffusion of Rural Innovations: Some Analytical Issues and Case Study of Improved Woodburning Stoves." *World Development* 11: 359–76.

———. 1986. *Cold Hearths and Barren Slopes: Woodfuel Crisis in the Third World*. New Delhi: Allied Press.

Ahmed, K., Y. A. Awe, D. F. Barnes, M. L. Cropper, and M. Kojima. 2005. *Environmental Health and Traditional Fuel Use in Guatemala*. Washington, DC: World Bank.

Ahmed, N., M. Zeitlin, A. Beiser, C. Super, and S. Gershoff. 1993. "A Longitudinal Study of the Impact of Behavioural Change Intervention on Cleanliness, Diarrhoeal Morbidity, and Growth of Children in Rural Bangladesh." *Social Science and Medicine* 37 (2): 159–71.

Alam, D., G. Marks, A. Baqui, M. Yunus, and G. Fuchs. 2000. "Association between Clinical Type of Diarrhoea and Growth of Children under 5 Years in Rural Bangladesh." *International Journal of Epidemiology* 29 (5): 916–21.

Alderman, H., J. Behrman, V. Lavy, and R. Menon. 2001. "Child Health and School Enrollment: A Longitudinal Analysis." *Journal of Human Resources* 36 (1): 185–205.

Alderman, H., J. Konde-Lule, I. Sebuliba, D. Bundy, and A. Hall. 2006. "Effect on Weight Gain of Routinely Giving Albendazole to Preschool Children during Child Health Days in Uganda: Cluster Randomised Controlled Trial." *British Medical Journal* 333 (7559): 122.

Allen, S. J., A. Raiko, A. O'Donnell, N. D. E. Alexander, and J. B. Clegg. 1998. "Causes of Preterm Delivery and Intrauterine Growth Retardation in a Malaria Endemic Region of Papua New Guinea." *Archive of Disease in Childhood* 79 (2): F135–40.

Alvarado, B.-E., M. V. Zunzunegui, H. Delisle, and J. Osorno. 2005. "Growth Trajectories Are Influenced by Breast-Feeding and Infant Health in an Afro-Colombian Community." *Journal of Nutrition* 135 (9): 2171–78.

Awasthi, S., and D. Bundy. 2007. "Intestinal Nematode Infection and Anaemia in Developing Countries." *British Medical Journal* 334 (7603): 1065–66.

Awasthi, S., D. Bundy, and L. Savioli. 2003. "Helminthic Infections." *British Medical Journal* 327 (7412): 431–33.

Bairagi, R., M. K. Chowdhury, Y. J. Kim, G. T. Curlin, and R. H. Gray. 1987. "The Association between Malnutrition and Diarrhoea in Rural Bangladesh." *International Journal of Epidemiology* 16 (3): 477–81.

Banerjee, I., S. Ramani, B. Primrose, P. Moses, M. Iturriza-Gomara, J. J. Gray, S. Jaffar, M. Bindhu, J. P. Muliyil, D. W. Brown, M. K. Estes, and G. Kang. 2006. "Comparative Study of the Epidemiology of Rotavirus in Children from a Community-Based Birth Cohort and a Hospital in South India." *Journal of Clinical Microbiology* 44 (7): 2468–74.

Bang, A. T., R. A. Bang, S. B. Baitule, M. H. Reddy, and M. D. Deshmukh. 1999. "Effect of Home-Based Neonatal Care and Management of Sepsis on Neonatal Mortality: Field Trial in Rural India." *Lancet* 354 (9194): 1955–61.

Baumgartner, R., and E. Pollitt. 1983. "The Bacon Chow Study: Analyses of the Effects of Infectious Illness on Growth of Infants." *Nutrition Research* 3: 9–21.

Becker, S., R. E. Black, and K. H. Brown. 1991. "Relative Effects of Diarrhea, Fever, and Dietary Energy Intake on Weight Gain in Rural Bangladeshi Children." *American Journal of Clinical Nutrition* 53 (6): 1499–503.

Behrman, J., J. Hoddinott, J. Maluccio, E. Soler-Hampejsek, E. L. Behrman, R. Martorell, M. Ramirez-Zea, and A D. Stein. 2006. "What Determines Adult Cognitive Skills? Impacts of Pre-school, Schooling, and Post-schooling Experiences in Guatemala." PSC Working Paper, Population Studies Center, University of Pennsylvania, Philadelphia.

Bennett, J., M. Schooley, H. Traverso, S. B. Agha, and J. Boring. 1996. "Bundling, a Newly Identified Risk Factor for Neonatal Tetanus: Implications for Global Control." *International Journal of Epidemiology* 25 (4): 879–84.

Berg, R. 2005. "Bridging the Great Divide: Environmental Health and the Environmental Movement." *Journal of Environmental Health* 67 (6): 39–52.

Berkman, D. S., A. G. Lescano, R. H. Gilman, S. L. Lopez, and M. M. Black. 2002. "Effects of Stunting, Diarrhoeal Disease, and Parasitic Infection during Infancy on Cognition in Late Childhood: A Follow-up Study." *Lancet* 359 (9306): 564–71.

Bertozzi, S., N. Padian, J. Wegbreit, L. DeMaria, B. Feldman, H. Gayle, J. Gold, R. Grant, and M. Isbell. 2006. "HIV/AIDS Prevention and Treatment." In *Disease Control Priorities in Developing Countries*, 2nd ed., ed. D. T. Jamison, J. G. Breman, A. R. Measham, G. Alleyne, M. Claeson, D. B. Evans, P. Jha, A. Mills, and P. Musgrove, 331–69. Washington, DC: World Bank; New York: Oxford University Press.

Biran, A., L. Smith, J. Lines, J. Ensink, and M. Cameron. 2007. "Smoke and Malaria: Are Interventions to Reduce Exposure to Indoor Air Pollution Likely to Increase Exposure to Mosquitoes?" *Transactions of the Royal Society of Tropical Medicine and Hygiene* 101 (11): 1065–71.

Black, R. E., K. H. Brown, and S. Becker. 1984. "Effects of Diarrhea Associated with Specific Enteropathogens on the Growth of Children in Rural Bangladesh." *Pediatrics* 73 (6): 799–805.

Blössner, M., and M. de Onis. 2005. *Malnutrition: Quantifying the Health Impact at National and Local Levels.* WHO Environmental Burden of Disease 12. Geneva: World Health Organization.

Boeston, K., P. Kolsky, and C. Hunt. 2007. "Improving Urban Water and Sanitation Services: Health, Access, and Boundaries." In *Scaling Urban Environmental Challenges: From Local to Global and Back,* ed. P. Marcotullio and G. McGranahan, 106–31. London: Earthscan.

Breman, J. G., M. S. Alilio, and A. Mills. 2004. "Conquering the Intolerable Burden of Malaria: What's New, What's Needed—A Summary." *American Journal of Tropical Medical Hygiene* 71 (2 Suppl.): 1–15.

Brenzel, L., L. J. Wolfson, J. Fox-Rushby, M. Miller, and N. A. Halsey. 2006. "Vaccine-Preventable Diseases." In *Disease Control Priorities in Developing Countries,* 2nd ed., ed. D. T. Jamison, J. G. Breman, A. R. Measham, G. Alleyne, M. Claeson, D. B. Evans, P. Jha, A. Mills, and P. Musgrove, 389–411. Washington, DC: World Bank; New York: Oxford University Press.

Briend, A., K. Hasan, K. Aziz, and B. Hoque. 1989. "Are Diarrhoea Control Programmes Likely to Reduce Childhood Malnutrition? Observations from Rural Bangladesh." *Lancet* 2 (8658): 319–22.

Briscoe, J. 1987. "The Author Replies." *American Journal of Epidemiology* 125 (5): 922–25.

Brown, K. H. 2003. "Symposium: Nutrition and Infection, Prologue and Progress since 1968—Diarrhea and Malnutrition." *Journal of Nutrition* 133 (1): 328S–32S.

Bruce, N., R. Perez-Padilla, and R. Albalak. 2002. "The Health Effects of Indoor Air Pollution Exposure in Developing Countries." WHO/SDE/OEH/02.05. World Health Organization, Geneva. http://whqlibdoc.who.int/hq/2002/WHO_SDE_OEH_02.05.pdf.

Brush, G., G. Harrison, and J. Waterlow. 1997. "Effects of Early Disease on Later Growth, and Early Growth on Later Disease, in Khartoum Infants." *Annals of Human Biology* 24 (3): 187–95.

Bryce, J., C. Boschi-Pinto, K. Shibuya, and R. Black. 2005. "WHO Estimates of the Causes of Death in Children." *Lancet* 365 (9465): 1147–52.

Burra, S., S. Patel, and T. Kerr. 2003. "Community Designed, Built and Managed Toilet Blocks in Indian Cities." *Environment and Urbanization* 15 (2): 11–32.

Cairncross, S. 1987. "Ingested Dose and Diarrhea Transmission Routes." *American Journal of Epidemiology* 125 (5): 921–22.

———. 2003. "Handwashing with Soap: A New Way to Prevent ARIs." *Tropical Medicine and International Health* 8 (8): 677–79.

Cairncross, S., U. Blumenthal, P. Kolsky, L. Moraes, and A. T. Tayeh. 1996. "The Public and Domestic Domains in the Transmission of Disease." *Tropical Medicine and International Health* 1 (1): 27–34.

Cairncross, S., and J. Kinnear. 1992. "Elasticity of Demand for Water in Khartoum, Sudan." *Social Science and Medicine* 34 (2): 183–89.

Cairncross, S., and P. Kolsky. 1997. "Water, Waste, and Well-Being: A Multicountry Study." *American Journal of Epidemiology* 146 (4): 359–61.

Cairncross, S., and V. Valdmanis. 2006. "Water Supply, Sanitation, and Hygiene Promotion." In *Disease Control Priorities in Developing Countries,* 2nd. ed., ed. D. T. Jamison, J. G. Breman,

A. R. Measham, G. Alleyne, M. Claeson, D. B. Evans, P. Jha, A. Mills, and P. Musgrove, 771–92. Washington, DC: World Bank; New York: Oxford University Press.

Campbell, D., M. Elia, and G. Lunn. 2003. "Growth Faltering in Rural Gambian Infants Is Associated with Impaired Small Intestinal Barrier Function, Leading to Endotoxemia and Systemic Infection." *Journal of Nutrition* 133 (5): 1332–38.

Campbell-Lendrum, D., C. Corvalán, and M. Neira. 2007. "Global Climate Change: Implications for International Public Health Policy." *Bulletin of the World Health Organization* 85 (3): 235–37.

Carson, Rachel. 1962. *Silent Spring*. New York: Houghton Mifflin.

Carter, R., and K. Mendis. 2006. "Measuring Malaria." *American Journal of Tropical Medical Hygiene* 74 (2): 187–88.

Cattand, P., P. Desjeux, M. Guzman, J. Jannin, A. Kroeger, A. Medici, P. Musgrove, M. B. Nathan, A. Shaw, and C. Schofield. 2006. "Tropical Diseases Lacking Adequate Control Measures: Dengue, Leishmaniasis, and African Trypanosomiasis." In *Disease Control Priorities in Developing Countries*, 2nd ed., ed. D. T. Jamison, J. G. Breman, A. R. Measham, G. Alleyne, M. Claeson, D. B. Evans, P. Jha, A. Mills, and P. Musgrove, 451–66. Washington, DC: World Bank; New York: Oxford University Press.

Caulfield, L. E., M. de Onis, M. Blössner, and R. E. Black. 2004. "Undernutrition as an Underlying Cause of Child Deaths Associated with Diarrhea, Pneumonia, Malaria, and Measles." *American Journal of Clinical Nutrition* 80 (1): 193–98.

Caulfield, L. E., S. A. Richard, J. A. Rivera, P. Musgrove, and R. E. Black. 2006. "Stunting, Wasting, and Micronutrient Deficiency Disorders." In *Disease Control Priorities in Developing Countries*, 2nd ed., ed. D. T. Jamison, J. G. Breman, A. R. Measham, G. Alleyne, M. Claeson, D. B. Evans, P. Jha, A. Mills, and P. Musgrove, 551–68. Washington, DC: World Bank; New York: Oxford University Press.

Chatterjee, N., and P. Hartge. 2003. "Apportioning Causes, Targeting Populations, and Predicting Risks: Population Attributable Fractions." *European Journal of Epidemiology* 18 (10): 933–35.

Checkley, W., L. D. Epstein, R. H. Gilman, R. E. Black, L. Cabrera, and C. R. Sterling. 1998. "Effects of *Cryptosporidium parvum* Infection in Peruvian Children: Growth Faltering and Subsequent Catch-up Growth." *American Journal of Epidemiology* 148 (5): 497–506.

Checkley, W., L. D. Epstein, R. H. Gilman, L. Cabrera, and R. E. Black. 2003. "Effects of Acute Diarrhea on Linear Growth in Peruvian Children." *American Journal of Epidemiology* 157 (2): 166–75.

Checkley, W., R. H. Gilman, L. D. Epstein, M. Suarez, J. F. Diaz, L. Cabrera, R. E. Black, and C. R. Sterling. 1997. "Asymptomatic and Symptomatic Cryptosporidiosis: Their Acute Effect on Weight Gain in Peruvian Children." *American Journal of Epidemiology* 145 (2): 156–63.

Chisti, M. J., M. I. Hossain, M. A. Malek, A. S. Faruque, T. Ahmed, and M. A. Salam. 2007. "Characteristics of Severely Malnourished Under-Five Children Hospitalized with Diarrhoea and Their Policy Implications." *Acta Paediatrica* 96 (5): 693–96.

Chu, D., R. D. Bungiro, M. Ibanez, L. M. Harrison, E. Campodonico, B. F. Jones, J. Mieszczanek, P. Kuzmic, and M. Cappello. 2004. "Molecular Characterization of *Ancylostoma ceylanicum* Kunitz-Type Serine Protease Inhibitor: Evidence for a Role in Hookworm-Associated Growth Delay." *Infection and Immunity* 72 (4): 2214–21.

Cohen, R. J., K. H. Brown, J. Canahuati, L. L. Rivera, and K. G. Dewey. 1995. "Determinants of Growth from Birth to 12 Months among Breast-Fed Honduran Infants in Relation to Age of Introduction of Complementary Foods." *Pediatrics* 96 (3): 504–10.

Condon-Paoloni, D., J. Cravioto, F. Johnston, E. De Licardie, and T. Scholl. 1977. "Morbidity and Growth of Infants and Young Children in a Rural Mexican Village." *American Journal of Public Health* 67 (7): 651–56.

Coovadia, H. M., N. C. Rollins, R. M. Bland, K. Little, A. Coutsoudis, M. L. Bennish, and M.-L. Newell. 2007. "Mother-to-Child Transmission of HIV-1 Infection during Exclusive Breastfeeding in the First 6 Months of Life: An Intervention Cohort Study." *Lancet* 369 (9567): 1107–16.

Crawley, J., J. Hill, J. Yartey, M. Robalo, A. Serufilira, A. Ba-Nguz, E. Roman, A. Palmer, K. Asamoa, and R. Steketee. 2007. "From Evidence to Action? Challenges to Policy Change and Programme Delivery for Malaria in Pregnancy." *Lancet Infectious Diseases* 7 (2): 145–55.

Cutler, D. M., and G. Miller. 2005. "The Role of Public Health Improvements in Health Advances: The Twentieth-Century United States." *Demography* 42 (1): 1–22.

Cutts, F., S. Zaman, G. Enwere, S. Jaffar, O. Levine, J. Okoko, C. Oluwalana, A. Vaughan, S. Obaro, A. Leach, K. McAdam, E. Biney, M. Saaka, U. Onwuchekwa, F. Yallop, N. Pierce, B. Greenwood, R. Adegbola, and the Gambian Pneumococcal Vaccine Trial Group. 2005. "Efficacy of Nine-Valent Pneumococcal Conjugate Vaccine against Pneumonia and Invasive Pneumococcal Disease in The Gambia: Randomised, Double-Blind, Placebo-Controlled Trial." *Lancet* 365 (9465): 1139–46.

Dai, D., and W. Walker. 1999. "Protective Nutrients and Bacterial Colonization in the Immature Human Gut." *Advances in Pediatrics* 46: 353–82.

Dandoy, S. 1997. "The State Public Health Department." In *Principles of Public Health Practice*, ed. F. D. Scutchfield and C. W. Keck, 83. Albany: Delmar.

Daniels, M. C., and L. S. Adair. 2004. "Growth in Young Filipino Children Predicts Schooling Trajectories through High School." *Journal of Nutrition* 134 (6): 1439–46.

Das Gupta, M., D. Ghosh, K. Datta, and S. Chakrabarti. 2006. "Harnessing Local Governments and Communities to Improve Environmental Sanitation and Public Health in India." Paper presented at the Workshop on Fiscal Decentralization and Local Governance in India, organized by the Indian Institute of Public Administration in partnership with the government of Japan and the World Bank, Goa, India, December 11–15, 2006.

Dasgupta, S., M. Huq, M. Khaliquzzaman, K. Pandey, and D. Wheeler. 2004a. *Indoor Air Quality for Poor Families: New Evidence from Bangladesh.* Washington, DC: World Bank.

———. 2004b. *Who Suffers from Indoor Air Pollution? Evidence from Bangladesh.* Washington, DC: World Bank.

Desai, M., S. Mehta, and K. Smith. 2004. *Indoor Smoke from Solid Fuels: Assessing the Environmental Burden of Disease at National and Local Levels.* Environmental Burden of Disease Series 4. Geneva: World Health Organization.

Dickson, R., S. Awasthi, P. Williamson, C. Demellweek, and P. Garner. 2000. "Effects of Treatment for Intestinal Helminth Infection on Growth and Cognitive Performance in Children: Systematic Review of Randomised Trials." *British Medical Journal* 320 (7251): 1697–701.

Eisenberg, J. N. S., W. Cevallos, K. Ponce, K. Levy, S. J. Bates, J. C. Scott, A. Hubbard, N. Vieira, P. Endara, M. Espinel, G. Trueba, L. W. Riley, and J. Trostle. 2006. "Environmental Change and Infectious Disease: How New Roads Affect the Transmission of Diarrheal Pathogens in Rural Ecuador." *Proceedings of the National Academy of Sciences* 103 (51): 19460–65.

Eisenberg, J. N. S., J. C. Scott, and T. Porco. 2007. "Integrating Disease Control Strategies: Balancing Water Sanitation and Hygiene Interventions to Reduce Diarrheal Disease Burden." *American Journal of Public Health* 97 (5): 846–52.

Eriksson, J. 2005. "The Fetal Origins Hypothesis: 10 Years On." *British Medical Journal* 330: 1096–97.

Ezzati, M., A. D. Lopez, A. Rodgers, and C. J. L. Murray, ed. 2004. *Comparative Quantification of Health Risks: Global and Regional Burden of Disease Attributable to Selected Major Risk Factors*. Geneva: World Health Organization.

Ezzati, M., A. Rodgers, A. D. Lopez, S. Vander Hoorn, and C. J. L. Murray. 2004. "Mortality and Burden of Disease Attributable to Individual Risk Factors." In *Comparative Quantification of Health Risks: Global and Regional Burden of Disease Attributable to Selected Major Risk Factors*, vol. 2, ed. M. Ezzati, A. D. Lopez, A. Rodgers, and C. J. L. Murray, 2141–65. Geneva: World Health Organization.

Ezzati, M., S. Vander Hoorn, A. Rodgers, A. D. Lopez, C. D. Mathers, and C. J. L. Murray. 2004. "Potential Health Gains from Reducing Multiple Risk Factors." In *Comparative Quantification of Health Risks: Global and Regional Burden of Disease Attributable to Selected Major Risk Factors*, vol. 2, ed. M. Ezzati, A. D. Lopez, A. Rodgers, and C. J. L. Murray, 2167–90. Geneva: World Health Organization.

Ezzati, M., S. Vander Hoorn, A. Rodgers, A. D. Lopez, C. D. Mathers, C. J. L. Murray, and the Comparative Risk Assessment Collaborating Group. 2003. "Estimates of Global and Regional Potential Health Gains from Reducing Multiple Major Risk Factors." *Lancet* 362 (9380): 271–80.

FAO (Food and Agriculture Organization). 2006. *The State of Food Insecurity in the World 2006: Eradicating World Hunger—Taking Stock Ten Years after the World Food Summit*. Rome: FAO.

Favin, M. 2004. "Promoting Hygiene Behavior Change within C-IMCI: The Peru and Nicaragua Experience." Activity Report 143, Environmental Health Project, Office of Health, Infectious Diseases, and Nutrition, Bureau for Global Health, U.S. Agency for International Development, Washington, DC. http://www.ehproject.org/PDF/Activity_Reports/AR-143CIMCI%20-format.pdf.

Favin, M., M. Yacoob, and D. Bendahmane. 1999. "Behavior First: A Minimum Package of Environmental Health Behaviors to Improve Child Health." Applied Study 10, Environmental Health Project, Office of Health, Infectious Diseases, and Nutrition, Bureau for Global Programs, U.S. Agency for International Development, Washington, DC. http://pdf.usaid.gov/pdf_docs/PNACF961.pdf.

Federal Bureau of Statistics. 2005. *Pakistan Social and Living Standards Measurement Survey 2004–05*. Islamabad: Federal Bureau of Statistics, Government of Pakistan.

Fegan, G. W., A. M. Noor, W. S. Akhwale, S. Cousens, and R. W. Snow. 2007. "Effect of Expanded Insecticide-Treated Bednet Coverage on Child Survival in Rural Kenya: A Longitudinal Study." *Lancet* 370 (9592): 1035–39.

Ferrie, J., and W. Troesken. 2004. "Death in the City: Mortality and Access to Public Water and Sewer in Chicago, 1880." Northwestern University, Evanston, IL.

Fewtrell, L., A. Prüss-Üstün, R. Bos, F. Gore, and J. Bartram. 2007. *Water, Sanitation, and Hygiene: Quantifying the Health Impact at National and Local Levels in Countries with Incomplete Water Supply and Sanitation Coverage*. Environmental Burden of Disease Series 15. Geneva: World Health Organization.

Fikree, F., M. Rahbar, and H. Berendes. 2000. "Risk Factors for Stunting and Wasting at Age Six, Twelve, and Twenty-Four Months for Squatter Children of Karachi, Pakistan." *Journal of the Pakistan Medical Association* 50 (10): 341–48.

Fishman, S., L. E. Caulfield, M. de Onis, M. Blössner, A. Hyder, L. Mullany, and R. E. Black. 2004. "Childhood and Maternal Underweight." In *Comparative Quantification of Health*

Risks: Global and Regional Burden of Disease Attributable to Selected Major Risk Factors, vol. 1, ed. M. Ezzati A. D. Lopez, A. Rodgers, and C. J. L. Murray, 39–61. Geneva: World Health Organization.

Flemming, K., W. van der Hoek, F. P. Amerasinghe, C. Muteroc, and E. Boelee. 2004. "Engineering and Malaria Control: Learning from the Past 100 Years." *Acta Tropica* 89 (2): 99–108.

Frenk, J. 2006. "Bridging the Divide: Global Lessons from Evidence-Based Health Policy in Mexico." *Lancet* 368 (9539): 954–61.

Frias, J., and N. Mukherjee. 2005. *Private Sector Sanitation Delivery in Vietnam: Harnessing Market Power for Rural Sanitation.* Jakarta: World Bank.

Galiani, S., P. Gertler, and E. Schargrodsky. 2005. "Water for Life: The Impact of the Privatization of Water Services on Child Mortality." *Journal of Political Economy* 113 (1): 83–120.

Ghana Statistical Service. 2003. *The Ghana Demographic and Health Survey 2003.* Accra: Ghana Statistical Service. http://www.measuredhs.com/aboutsurveys/search/metadata.cfm? surv_id=235&ctry_id=14&SrvyTp=type.

Glewwe, P., and H. Jacoby. 1995. "An Economic Analysis of Delayed Primary School Enrollment in a Low-Income Country: The Role of Early Childhood Nutrition." *Review of Economics and Statistics* 77 (1): 156–69.

Glewwe, P., H. Jacoby, and E. King. 2001. "Early Childhood Nutrition and Academic Achievement: A Longitudinal Analysis." *Journal of Public Economics* 81 (3): 345–68.

Gragnolati, M., M. Shekar, M. Das Gupta, C. Bredenkamp, and Y.-K. Lee. 2006. *India's Malnourished Children: A Call for Reform and Action.* Washington, DC: World Bank.

Graham, W., J. Cairns, S. Bhattacharya, C. H. W. Bullough, Z. Quayyum, and K. Rogo. 2006. "Maternal and Perinatal Conditions." In *Disease Control Priorities in Developing Countries,* 2nd ed., ed. D. T. Jamison, J. G. Breman, A. R. Measham, G. Alleyne, M. Claeson, D. B. Evans, P. Jha, A. Mills, and P. Musgrove, 499–529. Washington, DC: World Bank; New York: Oxford University Press.

Grantham-McGregor, S., Y. B. Cheung, S. Cueto, P. Glewwe, L. Richter, and B. Strupp. 2007. "Child Development in Developing Countries: Developmental Potential in the First 5 Years for Children in Developing Countries." *Lancet* 369 (9555): 60–70.

Griffiths, M., and J. S. McGuire. 2005. "A New Dimension for Health Reform: The Integrated Community Child Health Program in Honduras." In *Health System Innovations in Central America: Lessons and Impact of New Approaches,* ed. G. La Forgia, 173–96. Washington, DC: World Bank.

Guerrant, D. I., S. R. Moore, A. A. M. Lima, P. D. Patrick, J. B. Schorling, and R. L. Guerrant. 1999. "Association of Early Childhood Diarrhea and Cryptosporidiosis with Impaired Physical Fitness and Cognitive Function Four–Seven Years Later in a Poor Urban Community in Northeast Brazil." *American Journal of Tropical Medicine and Hygiene* 61 (5): 707–13.

Guerrant, R. L., L. V. Kirchhoff, D. S. Shields, M. K. Nations, J. Leslie, M. A. De Sousa, J. G. Araujo, L. L. Correia, K. T. Sauer, K. E. McClelland, F. L. Trowbridge, and J. M. Hughes. 1983. "Prospective Study of Diarrheal Illnesses in Northeastern Brazil: Patterns of Disease, Nutritional Impact, Etiologies, and Risk Factors." *Journal of Infectious Diseases* 148 (6): 986–97.

Guerrant, R. L., A. A. M. Lima, and F. Davidson. 2000. "Micronutrients and Infection: Interactions and Implications with Enteric and Other Infections and Future Priorities." *Journal of Infectious Diseases* 182 (1 Suppl.): S134–38.

Guerrant, R. L., J. B. Schorling, J. F. McAuliffe, and M. A. D. Souza. 1992. "Diarrhea as a Cause and Effect of Malnutrition: Diarrhea Prevents Catch-up Growth and Malnutrition Increases Diarrhea Frequency and Duration." *American Journal of Tropical Medicine and Hygiene* 47 (1): 28–35.

Guyatt, H. L., and R. W. Snow. 2004. "Impact of Malaria during Pregnancy on Low Birth Weight in Sub-Saharan Africa." *Clinical Microbiology Reviews* 17 (4): 760–69.

Ha, D. Q., and T. Q. Huan. 1997. "Dengue Activity in Viet Nam and Its Control Programme, 1997–98." *Dengue Bulletin* 21 (December): 35–40. http://www.searo.who.int/en/Section10/Section332/Section519_2384.htm.

Habicht, J.-P., J. DaVanzo, and W. P. Butz. 1988. "Mother's Milk and Sewage: Their Interactive Effects on Infant Mortality." *Pediatrics* 81 (3): 456–61.

Hadi, H., M. Dibley, and K. West. 2004. "Complex Interactions with Infection and Diet May Explain Seasonal Growth Responses to Vitamin A in Preschool Aged Indonesian Children." *European Journal of Clinical Nutrition* 58 (7): 990–99.

Hadi, H., R. Stoltzfus, L. Moulton, M. Dibley, and K. West Jr. 1999. "Respiratory Infections Reduce the Growth Response to Vitamin A Supplementation in a Randomized Controlled Trial." *International Journal of Epidemiology* 28 (5): 874–81.

Harrison, G., G. Brush, and F. Zumrawi. 1993. "Motherhood and Infant Health in Khartoum." *Bulletin of the World Health Organization* 71 (5): 529–33.

Heaver, R., and Y. Kachondam. 2002. *Thailand's National Nutrition Program: Lessons in Management and Capacity Development.* Washington, DC: World Bank.

Henry, F. 1981. "Environmental Sanitation Infection and Nutritional Status of Infants in Rural St. Lucia, West Indies." *Transactions of the Royal Society of Tropical Medicine and Hygiene* 75 (4): 507–13.

Henry, F., N. Alam, K. Aziz, and M. Rahaman. 1987. "Dysentery, Not Watery Diarrhoea, Is Associated with Stunting in Bangladeshi Children." *Human Nutrition and Clinical Nutrition* 41 (4): 243–49.

Holveck, J., J. Ehrenberg, S. Ault, R. Rojas, J. Vasquez, M. Cerqueira, J. Ippolito-Shepherd, M. Genovese, and M. Periago. 2007. "Prevention, Control, and Elimination of Neglected Diseases in the Americas: Pathways to Integrated, Inter-Programmatic, Inter-Sectoral Action for Health and Development." *BMC Public Health* 7 (1): 6.

Horton, S. 1999. "Opportunities for Investments in Nutrition in Low-Income Asia." *Asian Development Review,* 17 (1–2): 246–73. http://www.adb.org/documents/periodicals/adr/pdf/ADR-Vol17-Horton.pdf.

Hotez, P. J., D. H. Molyneux, A. Fenwick, E. Ottesen, S. E. Sachs, and J. D. Sachs. 2006. "Incorporating a Rapid-Impact Package for Neglected Tropical Diseases with Programs for HIV/AIDS, Tuberculosis, and Malaria: A Comprehensive Pro-Poor Health Policy and Strategy for the Developing World." *PLoS Medicine* 3 (5): 576–84.

Howard, G., and J. Bartram. 2003. "Domestic Water Quantity, Service Level, and Health." WHO/SDE/WSH/03.02. World Health Organization, Geneva. http://www.who.int/water_sanitation_health/diseases/WSH0302exsum.pdf.

Hunt, Caroline. 2006. "Sanitation and Human Development." Occasional Paper 26, Human Development Report Office, United Nations Development Programme, New York.

Hunter, J. M. 1992. "Elephantiasis: A Disease of Development in North East Ghana." *Social Science and Medicine* 35 (5): 627–49.

———. 2003. "Inherited Burden of Disease: Agricultural Dams and the Persistence of Bloody Urine (*Schistosomiasis hematobium*) in the Upper East Region of Ghana, 1959–1997." *Social Science and Medicine* 56 (2): 219–34.

Hutton, G., and L. Haller. 2004. *Evaluation of the Cost and Benefits of Water and Sanitation Improvements at the Global Level.* WHO/SDE/WSH/04.04. Geneva: World Health Organization. http://www.who.int/water_sanitation_health/wsh0404/en/index.html.

Iannotti, L. L., J. M. Tielsch, M. M. Black, and R. E. Black. 2006. "Iron Supplementation in Early Childhood: Health Benefits and Risks." *American Journal Clinical Nutrition* 84 (6): 1261–76.

IPCC (Intergovernmental Panel on Climate Change). 2007. "Climate Change 2007: Impacts, Adaptation, and Vulnerability." In *Contribution of Working Group II to the Fourth Assessment Report of the Intergovernmental Panel on Climate Change*, ed. M. L. Parry, O. F. Canziani, J. P. Palutikof, P. J. van der Linden, and C. E. Hanson, 79–131. Cambridge, U.K.: Cambridge University Press.

Jacoby, H., and L. Wang. 2004. "Environmental Determinants of Child Mortality in Rural China: A Competing Risks Approach." Policy Research Working Paper 3241, World Bank, Washington, DC.

Jefferson, T., R. Foxlee, C. Del Mar, L. Dooley, E. Ferroni, B. Hewak, A. Prabhala, S. Nair, and A. Rivetti. 2007. "Interventions for the Interruption or Reduction of the Spread of Respiratory Viruses (Review)." *Cochrane Database of Systematic Reviews* (4).

Kay, B., and V. Nam. 2005. "New Strategy for Dengue Control in Vietnam." *Lancet* 365 (9459): 613–17.

Keiser, J., B. Singer, and J. Utzinger. 2005. "Reducing the Burden of Malaria in Different Eco-Epidemiological Settings with Environmental Management: A Systematic Review." *Lancet Infectious Diseases* 5 (11): 695–708.

Keusch, G. 2003. "The History of Nutrition: Malnutrition, Infection and Immunity." *Journal of Nutrition* 133 (1 Suppl.): 336S–40S.

Kimani-Murage, E. W., and A. Ngindu. 2007. "Quality of Water the Slum Dwellers Use: The Case of a Kenyan Slum." *Journal of Urban Health* 84 (6): 829–38.

Kivi, M., and Y. Tindberg. 2006. "*Helicobacter pylori* Occurrence and Transmission: A Family Affair?" *Scandinavian Journal of Infectious Diseases* 38 (6): 407–17.

Knapp, A. 2006. "Mainstreaming Hygiene and Sanitation into the Health Sector: Experience from Local 'at-Scale Sanitary Revolutions' and Recent Sector Developments in Ethiopia." Paper presented by W. Gebreselassie, Ethiopia Ministry of Health, at Water Week 2007, February 27–March 2. http://siteresources.worldbank.org/INTWRD/Resources/Worku_Gebreselassie_Ethiopia_Mainstreaming_Hygiene_and_Sanitation_Into_Preventive_Health_Care_Programs.pdf.

Kolsky, P., and U. Blumenthal. 1995. "Environmental Health Indicators and Sanitation-Related Disease in Developing Countries: Limitations to the Use of Routine Data Sources." *World Health Statistics Quarterly* 48 (2): 132–39.

Kolsky, P., E. Perez, W. Vandersypen, and L. Jensen. 2005. "Sanitation and Hygiene at the World Bank." Working Note No. 6, October 6, 2005. http://www-wds.worldbank.org/external/default/WDSContentServer/WDSP/IB/2005/11/30/000090341_20051130115025/Rendered/PDF/344690Sanitation0and0hygiene0at0wb.pdf.

Kolsteren, P., J. Kusin, and S. Kardjati. 1997. "Morbidity and Growth Performance of Infants in Madura, Indonesia." *Annales Tropical Pediatrics* 17 (3): 201–8.

Komisar, J. L. 2007. "Malaria Vaccines." *Frontiers in Bioscience* 12 (May): 3928–55.

Kosek, M., C. Bern, and R. L. Guerrant. 2003. "The Global Burden of Diarrhoeal Disease, as Estimated from Studies Published between 1992 and 2000." *Bulletin of the World Health Organization* 81 (3): 197–204.

Kotchian, S. 1997. "Perspectives on the Place of Environmental Health and Protection in Public Health and Public Health Agencies." *Annual Review of Public Health* 18: 245–59.

Kristensen, P. 2005. "Sub-Saharan Africa Region." *Environment Matters at the World Bank: Annual Review 2005*, 32–35.

Krupnick, A., B. Larsen, and E. Strukova. 2006. "Cost of Environmental Degradation in Pakistan: An Analysis of Physical and Monetary Losses in Environmental Health and Natural Resources." Background paper, World Bank, Washington, DC.

Kumie, A., and A. Ahmed. 2005. "An Overview of Environmental Health Status in Ethiopia with Particular Emphasis to Its Organization, Drinking Water, and Sanitation: A Literature Survey." *Ethiopia Journal of Health Development* 19 (2): 89–103.

Langhans, W. 2000. "Anorexia of Infection: Current Prospects." *Nutrition Research* 16 (10): 996–1005.

Larsen, B. 2006. "Ghana: Cost of Environmental Damage—An Analysis of Environmental Health." Background paper, World Bank, Washington, DC.

Laxminarayan, R., J. Chow, and S. Shahid-Salles. 2006. "Intervention Cost-Effectiveness Overview of Main Messages." In *Disease Control Priorities in Developing Countries*, 2nd ed., ed. D. T. Jamison, J. G. Breman, A. R. Measham, G. Alleyne, M. Claeson, D. B. Evans, P. Jha, A. Mills, and P. Musgrove, 35–86. Washington, DC: World Bank; New York: Oxford University Press.

Lim, M. 2001. "A Perspective on Tropical Sprue." *Current Gastroenterology Reports* 3 (4): 322–27.

Lima, A., S. Moore, M. Barboza, A. Soares, M. Schleupner, R. Newman, C. Sears, J. Nataro, D. Fedorko, T. Wuhib, J. Schorling, and R. L. Guerrant. 2000. "Persistent Diarrhea Signals a Critical Period of Increased Diarrhea Burdens and Nutritional Shortfalls: A Prospective Cohort Study among Children in Northeastern Brazil." *Journal of Infectious Diseases* 181 (5): 1643–51.

Listorti, J., and F. Doumani. 2001. *Environmental Health: Bridging the Gaps*. Washington, DC: World Bank.

Lorntz, B., A. M. Soares, S. R. Moore, R. Pinkerton, B. Gansneder, V. E. Bovbjerg, H. Guyatt, A. M. Lima, and R. L. Guerrant. 2006. "Early Childhood Diarrhea Predicts Impaired School Performance." *Pediatric Infectious Disease Journal* 25 (6): 513–20.

Luby, S., M. Agboatwalla, D. Feikin, J. Painter, W. Billhimer, and R. Hoekstra. 2005. "Effect of Handwashing on Child Health: A Randomised Controlled Trial." *Lancet* 366 (9481): 225–33.

Lucas, R. M., and A. J. McMichael. 2005. "Association or Causation? Evaluating Links between 'Environment and Disease.'" *Bulletin of the World Health Organization* 83 (10): 792–95.

Lunn, P., C. Northrop-Clewes, and R. Downes. 1991. "Intestinal Permeability, Mucosal Injury, and Growth Faltering in Gambian Infants." *Lancet* 338 (8772): 907–10.

Lutter, C., J. Mora, J. Habicht, K. Rasmussen, D. S. Robson, S. Sellers, C. Super, and M. Herrera. 1989. "Nutritional Supplementation: Effects on Child Stunting because of Diarrhea." *American Journal of Clinical Nutrition* 50 (1): 1–8.

Lvovsky, K. 2001. *Health and Environment*. Washington, DC: World Bank.

Mabaso, M. L. H., B. Sharp, and C. Lengeler. 2004. "Historical Review of Malarial Control in Southern Africa with Emphasis on the Use of Indoor Residual House-Spraying." *Tropical Medicine and International Health* 9 (8): 846–56.

Mahendradhata, Y., and F. Moerman. 2004. "Integration and Disease Control: Notes from the Prince Leopold Institute of Tropical Medicine Colloquium 2002." *Tropical Medicine and International Health* 9 (6): A5–10.

Majumder, R. 2004. "Total Sanitation Campaign: Changing the Face of Rural Burdwan." In *India Infrastructure Report 2004: Ensuring Value for Money*, ed. S. Morris, 325–26. New Delhi: Oxford University Press.

Malaviya, V. S., R. Kant, C. S. Pant, H. C. Srivastava, and R. S. Yadav. 2006. "Community Based Integrated Malaria Control with Reference to Involvement of Social Forestry Activities: An Experience." *Indian Journal of Community Medicine* 31 (4): 234–36.

Maleta, K., S. M. Virtanen, M. Espo, T. Kulmala, and P. Ashorn. 2003. "Childhood Malnutrition and Its Predictors in Rural Malawi." *Paediatric and Perinatal Epidemiology* 17 (4): 384–90.

Maluccio, J., J. Hoddinott, and J. Behrman. 2006. "The Impact of an Experimental Nutrition Intervention in Childhood on Education among Guatemalan Adults." IFPRI and Food Consumption and Nutrition Division Discussion Paper, International Food Policy Research Institute, Washington, DC.

Marodi, L. 2006. "Innate Cellular Immune Responses in Newborns." *Clinical Immunology* 118 (2–3): 137–44.

Martorell, R. 1995. "Results and Implications of the INCAP Follow-up Study." *Journal of Nutrition* 125 (4 Suppl.): 1127S–38S.

Martorell, R., J. P. Habicht, C. Yarbrough, A. Lechtig, R. E. Klein, and K. A. Western. 1975. "Acute Morbidity and Physical Growth in Rural Guatemalan Children." *American Journal of Disabled Children* 129 (11): 1296–301.

Martorell, R., C. Yarbrough, A. Lechtig, J. Habicht, and R. Klein. 1975. "Diarrheal Diseases and Growth Retardation in Preschool Guatemalan Children." *American Journal of Physical Anthropology* 43 (3): 341–46.

Mason, J., D. Sanders, P. Musgrove, and R. Galloway. 2006. "Community Health and Nutrition Programs." In *Disease Control Priorities in Developing Countries*, 2nd ed., ed. D. T. Jamison, J. G. Breman, A. R. Measham, G. Alleyne, M. Claeson, D. B. Evans, P. Jha, A. Mills, and P. Musgrove, 1053–76. Washington, DC: World Bank; New York: Oxford University Press.

Mata, L. 1992. "Diarrheal Disease as a Cause of Malnutrition." *American Journal of Tropical Medicine and Hygiene* 47 (1 Suppl.): 16–27.

Mathee, A., F. Swanepoel, and A. Swart. 1999. "Environmental Health Services." In *The South African Health Review*, ed. N. Crisp and A. Ntuli, 277–88. Durban, South Africa: Health Systems Trust. http://www.healthlink.org.za/uploads/files/chapter20_99.pdf.

Mavalankar, D., and S. Manjunath. 2004. "Sanitation and Water Supply: The Forgotten Infrastructure." In *India Infrastructure Report 2004: Ensuring Value for Money*, ed. S. Morris, 314–24. New Delhi: Oxford University Press.

Menendez, C., A. F. Fleming, and P. L. Alonso. 2000. "Malaria-Related Anaemia." *Parasitology Today* 16 (11): 469–76.

Modi, V., S. McDade, D. Lallement, and J. Saghir. 2005. "Energy Services for the Millennium Development Goals." Paper prepared for the Millennium Project of the United Nations Development Programme and World Bank, Washington, DC.

Molbak, K., M. Andersen, P. Aaby, N. Hojlyng, M. Jakobsen, M. Sodeman, and A. P. J. d. Silva. 1997. "Cryptosporidium Infection in Infancy as a Cause of Malnutrition: A Community Study from Guinea-Bissau, West Africa." *American Journal Clinical Nutrition* 65 (1): 149–52.

Moore, S., A. Lima, M. Conaway, J. Schorling, A. Soares, and R. L. Guerrant. 2001. "Early Childhood Diarrhoea and Helminthiases Associate with Long-Term Linear Growth Faltering." *International Journal of Epidemiology* 30 (6): 1457–64.

Morley, D., J. Bicknell, and M. Woodland. 1968. "Factors Influencing the Growth and Nutritional Status of Infants and Young Children in a Nigerian Village." *Transactions of the Royal Society of Tropical Medicine and Hygiene* 62 (2): 164–99.

Moser Jones, M. 2005. *Protecting Public Health in New York City: 200 Years of Leadership (1805–2005)*. New York: New York City Department of Health and Mental Hygiene.

Moy, R., T. Marshall, R. Choto, A. McNeish, and I. Booth. 1994. "Diarrhoea and Growth Faltering in Rural Zimbabwe." *European Journal of Clinical Nutrition* 48 (11): 810–21.

Mullany, L. C., G. L. Darmstadt, S. K. Khatry, J. Katz, S. C. LeClerg, S. Shrestra, R. Adhikari, and J. M. Tielsch. 2006. "Topical Applications of Chlorhexidine to the Umbilical Cord for Prevention of Omphalitis and Neonatal Mortality in Southern Nepal: A Community-Based, Cluster-Randomized Trial." *Lancet* 367 (9514): 910–18.

Murphy, H., B. Stanton, and J. Galbraith. 1997. *Prevention: Environmental Health Interventions to Sustain Child Survival*. Washington, DC: U.S. Agency for International Development.

Murray, C., T. Laakso, K. Shibuya, T. Laakso, K. Hill, and A. D. Lopez. 2007. "Can We Achieve Millennium Development Goal 4? New Analysis of Country Trends and Forecasts of under-5 Mortality to 2015." *Lancet* 370 (9592): 1040–54.

Nair, N., H. Gans, L. Lew-Yasukawa, A. C. Long-Wagar, A. Arvin, and D. E. Griffin. 2007. "Age-Dependent Differences in IgG Isotype and Avidity Induced by Measles Vaccine Received during the First Year of Life." *Journal of Infectious Diseases* 196 (9): 1339–45.

National Center for Infectious Diseases. 2006. "Malaria: Vector Control." Centers for Disease Control and Prevention, Atlanta. http://www.cdc.gov/malaria/control_prevention/vector_control.htm.

National Institute of Population Studies. 1992. *Pakistan Demographic and Health Survey 1990/91*. Islamabad: National Institute of Population Studies.

Nguyen, T. V., P. Le Van, C. Le Huy, K. N. Gia, and A. Weintraub. 2005. "Detection and Characterization of Diarrheagenic *Escherichia coli* from Young Children in Hanoi, Vietnam." *Journal of Clinical Microbiology* 43(2): 755–60.

Niehaus, M. D., S. R. Moore, P. D. Patrick, L. L. Derr, B. Lorntz, A. A. Lima, and R. L. Guerrant. 2002. "Early Childhood Diarrhea Is Associated with Diminished Cognitive Function 4 to 7 Years Later in Children in a Northeast Brazilian Shantytown." *American Journal of Tropical Medicine and Hygiene* 65 (5): 590–93.

Parkin, D. M. 2006. "The Global Health Burden of Infection-Associated Cancers in the Year 2002." *International Journal of Cancer* 118 (12): 3030–44.

Patrick, P. D., R. B. Oriá, V. Madhavan, R. C. Pinkerton, B. Lorntz, A. A. M. Lima, and R. L. Guerrant. 2005. "Limitations in Verbal Fluency Following Heavy Burdens of Early Childhood Diarrhea in Brazilian Shantytown Children." *Child Neuropsychology* 11 (3): 233–44.

Pearce, D. W., and D. Ulph. 1999. "A Social Discount Rate for the United Kingdom." In *Environmental Economics: Essays in Ecological Economics and Sustainable Development*, ed. D. W. Pearce, 268–85. Cheltenham, U.K.: Edward Elgar.

Pelletier, D. L. 1994. "The Relationship between Child Anthropometry and Mortality in Developing Countries: Implications for Policy, Programs, and Future Research." *Journal of Nutrition* 124 (10 Suppl.): 2047S–81S.

Pelletier, D. L., and E. A. Frongillo. 2003. "Changes in Child Survival Are Strongly Associated with Changes in Malnutrition in Developing Countries." *Journal of Nutrition* 133 (1): 107–19.

Pelletier, D. L., E. A. Frongillo, D. G. Schroeder, and J.-P. Habicht. 1994. "A Methodology for Estimating the Contribution of Malnutrition to Child Mortality in Developing Countries." *Journal of Nutrition* 124 (10 Suppl.): 2106S–22S.

PIDE (Pakistan Institute of Development Economics, Micronutrient Laboratories Aga Khan University, Medical Centre). 2003. *National Nutrition Survey 2001–2002*. Islamabad: Government of Pakistan, Planning Commission.

Poverty Environment Partnership. 2008. "Poverty, Health, and Environment: Placing Environmental Health on Countries' Development Agendas." Joint agency paper, World Bank, Washington, DC.

Powanda, M. C., and W. R. Beisel. 2003. "Metabolic Effects of Infection on Protein and Energy Status." *Journal of Nutrition* 133 (1 Suppl.): 322S–27S.

Prado, M., S. Cairncross, M. Barreto, A. Oliviera-Assis, and S. Rego. 2005. "Asymptomatic Giardiasis and Growth in Young Children: A Longitudinal Study in Salvador, Brazil." *Parasitology* 131 (1): 51–56.

Prüss-Üstün, A., and C. Corvalán. 2006. *Preventing Disease through Healthy Environments: Towards an Estimate of the Environmental Burden of Disease*. Geneva: World Health Organization.

Psacharopoulos, G., and H. Patrinos. 2004. "Returns to Investment in Education: A Further Update." *Education Economics* 12 (2): 111–34.

Public-Private Partnership for Handwashing with Soap. 2005. "Peru." World Bank, Washington, DC. http://www.globalhandwashing.org/Country%20act/Peru.htm.

Quddus, A., S. Luby, M. Rahbar, and Y. Pervaiz. 2002. "Neonatal Tetanus: Mortality Rate and Risk Factors in Loralai District, Pakistan." *International Journal of Epidemiology* 31 (3): 648–53.

Rabie, T., and V. Curtis. 2006. "Handwashing and Risk of Respiratory Infections: A Quantitative Systematic Review." *Tropical Medicine and International Health* 11 (3): 258–67.

Ramalingaswami, V., U. Jonsson, and J. Rohde. 1996. "Commentary: The Asian Enigma." In *The Progress of Nations,* ed. P. Adamson, 11–17. New York: United Nations Children's Fund.

RBM (Roll Back Malaria Partnership). 2005. *World Malaria Report 2005*. Geneva: United Nations Children's Fund and World Health Organization. http://www.rbm.who.int/wmr2005/index.html.

Rehfuess, E., S. Mehta, and A. Prüss-Üstün. 2006. "Assessing Household Solid Fuel Use: Multiple Implications for the Millennium Development Goals." *Environmental Health Perspectives* 114 (3): 373–78.

Remme, J. H. F., P. Feentstra, P. R. Lever, A. Medici, C. Morel, M. Noma, K. D. Ramaiah, F. Richards, A. Seketeli, G. Schmunis, W. H. van Brakel, and A. Vassall. 2006. "Tropical Diseases Targeted for Elimination: Chagas Disease, Lymphatic Filariasis, Onchocerciasis, and Leprosy." In *Disease Control Priorities in Developing Countries*, 2nd ed., ed. D. T. Jamison, J. G. Breman, A. R. Measham, G. Alleyne, M. Claeson, D. B. Evans, P. Jha, A. Mills, and P. Musgrove, 433–49. Washington, DC: World Bank; New York: Oxford University Press.

Rosen, G. 1958. *A History of Public Health*. Baltimore, MD: Johns Hopkins University Press.

Rowland, M., T. J. Cole, and R. Whitehead. 1977. "A Quantitative Study into the Role of Infection in Determining Nutritional Status in Gambian Village Children." *Journal of Nutrition* 37 (3): 441–50.

Rowland, M., S. G. Rowland, and T. J. Cole. 1988. "Impact of Infection on the Growth of Children from 0 to 2 Years in an Urban West African Community." *American Journal of Clinical Nutrition* 47 (1): 134–38.

Sachdev, H. S., C. H. D. Fall, C. Osmond, R. Lakshmy, S. K. D. Biswas, S. D. Leary, K. S. Reddy, D. J. P. Barker, and S. K. Bhargava. 2005. "Anthropometric Indicators of Body

Composition in Young Adults: Relation to Size at Birth and Serial Measurements of Body Mass Index in Childhood in the New Delhi Birth Cohort." *American Journal of Clinical Nutrition* 82 (2): 456–66.

Sarraf, M., B. Larsen, and M. Owaygen. 2004 "Cost of Environmental Degradation: The Case of Lebanon and Tunisia." Environment Department Paper 97, World Bank, Washington, DC.

Sato, H., P. Albrecht, D. W. Reynolds, S. Stagno, and F. A. Ennis. 1979. "Transfer of Measles, Mumps, and Rubella Antibodies from Mother to Infant: Its Effect on Measles, Mumps, and Rubella Immunization." *American Journal of Diseases of Children* 133 (12): 1240–43.

Satterthwaite, D. 2007. "In Pursuit of a Healthy Urban Environment in Low- and Middle-Income Nations." In *Scaling Urban Environmental Challenges: From Local to Global and Back*, ed. P. Marcotullio and G. McGranahan, 69–105. London: Earthscan.

Schorling, J., J. F. Mcauliffe, M. A. De Souza, and R. L. Guerrant. 1990. "Malnutrition Is Associated with Increased Diarrhoea Incidence and Duration among Children in an Urban Brazilian Slum." *International Journal of Epidemiology* 19 (3): 728–35.

Scrimshaw, N. S. 2003. "Historical Concepts of Interactions, Synergism, and Antagonism between Nutrition and Infection." *Journal of Nutrition* 133 (1 Suppl.): 316S–21S.

Scrimshaw, N. S., C. E. Taylor, and J. E. Gordon. 1959. "Interactions of Nutrition and Infection." *American Journal of Medical Sciences* 237 (3): 367–72.

———. 1968. *Interactions of Nutrition and Infection*. Geneva: World Health Organization.

Sedgwick, W. T., and S. Macnutt. 1908. "An Examination of the Theorem of Allen Hazen That for Every Death from Typhoid Fever Avoided by the Purification of Public Water Supplies Two or Three Deaths Are Avoided from Other Causes." *Science* 28 (711): 215–16.

Sepúlveda, J., F. Bustreo, R. Tapia, J. A. Rivera, R. Lozano, G. Oláiz, V. Partida, L. García-García, and J. L. Valdespino. 2006. "Improvement of Child Survival in Mexico: The Diagonal Approach." *Lancet* 368 (9551): 2017–27.

Shekar, M., R. Heaver, Y.-K. Lee, and World Bank. 2006. *Repositioning Nutrition as Central to Development: A Strategy for Large-Scale Action*. Washington, DC: World Bank.

Shrimpton, R., C. Victora, M. de Onis, R. Costa Lima, M. Blössner, and G. Clugston. 2001. "The Worldwide Timing of Growth Faltering: Implications for Nutritional Interventions." *Pediatrics* 107 (5): e75.

Singleton, R., T. Hennessy, L. Bulkow, L. Hammitt, T. Zulz, D. Hurlburt, J. Butler, K. Rudolph, and A. Parkinson. 2007. "Invasive Pneumococcal Disease Caused by Nonvaccine Serotypes among Alaska Native Children with High Levels of 7-Valent Pneumococcal Conjugate Vaccine Coverage." *Journal of the American Medical Association* 297 (16): 1784–92.

Sinton, J., K. Smith, J. Peabody, Y. Liu, X. Zhang, and R. Edwards. 2004. "An Assessment of Programs to Promote Improved Household Stoves in China." *Energy for Sustainable Development* 8 (3): 33–52.

Smith, K. R., C. F. Corvalán, and T. Kjellström. 1999. "How Much Global Ill Health Is Attributable to Environmental Factors?" *Epidemiology and Infection* 10 (5): 573–84.

Smith, K. R., S. Mehta, and M. Maeuszahl-Feuz. 2004. "Indoor Air Pollution from Household Use of Solid Fuels." In *Comparative Quantification of Health Risks: Global and Regional Burden of Disease Attributable to Selected Major Risk Factors*, vol. 2, ed. M. Ezzati, A. D. Lopez, A. Rodgers, and C. J. L. Murray, 1435–93. Geneva: World Health Organization.

Snow, R., M. Craig, C. Newton, and R. Stekete. 2003. *The Public Health Burden of* Plasmodium falciparum *Malaria in Africa: Deriving the Numbers.* Bethesda, MD: Fogarty International Center, National Institutes of Health.

Steiner, T., A. Lima, J. Nataro, and R. L. Guerrant. 1998. "Enteroaggregative *Escherichia coli* Produce Intestinal Inflammation and Growth Impairment and Cause Interleukin-8 Release from Intestinal Epithelial Cells." *Journal of Infectious Diseases* 177 (1): 88–96.

Steketee, R. W. 2003. "Pregnancy, Nutrition and Parasitic Diseases." *Journal of Nutrition* 133 (5): 1661S–67S.

Stephensen, C. B. 1999. "Burden of Infection on Growth Failure." *Journal of Nutrition* 129 (2): 534–38.

Sur, D., D. Saha, B. Manna, K. Rajendran, and S. Bhattacharya. 2005. "Periodic Deworming with Albendazole and Its Impact on Growth Status and Diarrhoeal Incidence among Children in an Urban Slum of India." *Transactions of the Royal Society of Tropical Medicine and Hygiene* 99 (4): 261–67.

Svedberg, P. 2006. "Child Malnutrition in Shining India: A X-state Empirical Analysis." Institute for Economic Studies, Stockholm. http://www.bvsde.paho.org/bvsacd/cd56/svedberg-070906.pdf.

Taylor-Robinson, D., A. Jones, and P. Garner. 2007. "Deworming Drugs for Treating Soil-Transmitted Intestinal Worms in Children: Effects on Growth and School Performance." *Cochrane Database of Systematic Reviews* 17 (4).

Thapar, N., and I. Sanderson. 2004. "Diarrhoea in Children: An Interface between Developing and Developed Countries." *Lancet* 363 (9409): 641–53.

Thomas, E. P., J. R. Seager, and A. Mathee. 2002. "Environmental Health Challenges in South Africa: Policy Lessons from Case Studies." *Health and Place* 8 (4): 251–61.

Touré, Y. 2001. "Malaria Vector Control in Africa: Strategies and Challenges." Report from a symposium held at the 2001 American Association for the Advancement of Science Annual Meeting, San Francisco, February 17. http://www.aaas.org/international/africa/malaria/toure.html.

Trades Union Congress. 2004. "Incomes in Ghana: Policy Discussion Paper." Trades Union Congress, Accra.

UNICEF (United Nations Children's Fund). 2005. "Global Database on Low Birthweight." UNICEF, New York. http://www.childinfo.org/areas/birthweight/database.php.

———. 2006. "Water." UNICEF, New York. http://www.childinfo.org/areas/water/.

———. 2008. "Info by Country." UNICEF, New York. http://www.unicef.org/infobycountry/ghana_statistics.html.

———. 2008. "Info by Country." UNICEF, New York. http://www.unicef.org/infobycountry/pakistan_pakistan_statistics.html#47.

Valentiner-Branth, P., H. Steinsland, G. Santos, M. Perch, K. Begtrup, M. K. Bhan, F. Dias, P. Aaby, H. Sommerfelt, and K. Molbak. 2001. "Community-Based Controlled Trial of Dietary Management of Children with Persistent Diarrhea: Sustained Beneficial Effect on Ponderal and Linear Growth." *American Journal of Clinical Nutrition* 73 (5): 968–74.

VanDerslice, J., B. Popkin, and J. Briscoe. 1994. "Drinking-Water Quality, Sanitation, and Breast-Feeding: Their Interactive Effects on Infant Health." *Bulletin of the World Health Organization* 72 (4): 589–601.

van Geertruyden, J., F. Thomas, A. Erhart, and U. D'Alessandro. 2004. "The Contribution of Malaria in Pregnancy to Perinatal Mortality." *American Journal of Tropical Medicine and Hygiene* 71 (2 Suppl.): 35–40.

Vergnano, S., M. Sharland, P. Kazembe, C. Mwansambo, and P. T. Heath. 2005. "Neonatal Sepsis: An International Perspective." *Archives of Disease in Childhood: Fetal and Neonatal Edition* 90 (3): F220–24.

Victora, C. G., F. C. Barros, B. R. Kirkwood, and P. Vaughan. 1990. "Pneumonia, Diarrhea, and Growth in the First 4 Years of Life: A Longitudinal Study of 5,914 Urban Brazilian Children." *American Journal Clinical Nutrition* 52 (2): 391–96.

Victora, C. G., B. Fenn, J. Bryce, and B. R. Kirkwood. 2005. "Co-coverage of Preventive Interventions and Implications for Child-Survival Strategies: Evidence from National Surveys." *Lancet* 366 (9495): 1460–66.

Villamor, E., M. Fataki, R. Bosch, R. Mbise, and W. Fawzi. 2004. "Human Immunodeficiency Virus Infection, Diarrheal Disease, and Sociodemographic Predictors of Child Growth." *Acta Paediatrica* 93 (3): 372–79.

von Schirnding, Y. 2005. "Short Report: The World Summit on Sustainable Development: Reaffirming the Centrality of Health." *Globalization and Health* 1 (8): 1–6.

Vu, S. N., T. Y. Nguyen, T. V. Phong, T. U. Ninh, L. Q. Mai, L. V. Lo, L. T. Nghia, A. Bektas, A. Briscombe, J. G. Aaskov, P. A. Ryan, and B. H. Kay. 2005. "Elimination of Dengue by Community Programs Using Mesocyclops (*Copepoda*) against *Aedis Aegypti* in Central Vietnam." *American Journal of Tropical Medicine and Hygiene* 72 (1): 67–73.

Wagstaff, A., and M. Claeson. 2004. *The Millennium Development Goals for Health: Rising to the Challenges.* Washington, DC: World Bank.

Walker, S. P., S. M. Chang, C. A. Powell, and S. M. Grantham-McGregor. 2005. "Effects of Early Childhood Psychosocial Stimulation and Nutritional Supplementation on Cognition and Education in Growth-Stunted Jamaican Children: Prospective Cohort Study." *Lancet* 366 (9499): 1804–7.

Walker, S. P., T. D. Wachs, J. Meeks Garder, B. Lozoff, G. A. Wasserman, E. Pollitt, and J. Carter. 2007. "Child Development: Risk Factors for Adverse Outcomes in Developing Countries." *Lancet* 369 (9556): 145–57.

Watson-Jones, D., H. A. Weiss, J. M. Changalucha, J. Todd, B. Gumodoka, J. Bulmer, R. Balira, D. Ross, K. Mugeye, R. Hayes, and D. Mabey. 2007. "Adverse Birth Outcomes in United Republic of Tanzania: Impact and Prevention of Maternal Risk Factors." *Bulletin of the World Health Organization* 85 (1): 9–18.

West, K. P. 2003. "Vitamin A Deficiency Disorders in Children and Women." *Food and Nutrition Bulletin* 24 (4): S78–S90.

WHO (World Health Organization). 2000. *Addressing the Links between Indoor Air Pollution, Household Energy and Human Health.* Geneva: WHO.

———. 2002. *The World Health Report 2002: Reducing Risks, Promoting Healthy Life.* Geneva: WHO. http://www.who.int/whr/2002/en/index.html.

———. 2003. *Insecticide-Treated Mosquito Net Interventions: A Manual for National Control Programme Managers.* Geneva: WHO.

———. 2004. *World Health Report 2004.* Geneva: WHO.

———. 2006. *Fuel for Life: Household Energy and Health.* Geneva: WHO.

———. 2007a. "Burden of Disease Project." WHO, Geneva. http://www.who.int/healthinfo//bodproject/en/index.html.

———. 2007b. "Children's Environmental Health." WHO, Geneva. http://www.who.int/ceh/en/.

———. 2007c. *Combating Waterborne Disease at the Household Level.* Geneva: WHO.

———. 2007d. "Malaria." Fact Sheet 94, WHO, Geneva. http://www.who.int/mediacentre/factsheets/fs094/en/index.html.

————. 2007e. *World Health Report 2007— A Safer Future: Global Public Health Security in the 21st Century.* Geneva: WHO.

————. 2008. "Environmental Health." WHO, Geneva. http://www.who.int/topics/environmental_health/en/.

————. n.d. "Quantification of the Burden of Disease Attributable to Environmental Risk Factors." WHO, Geneva. http://www.who.int/quantifying_ehimpacts/summaryEBD.pdf.

WHO (World Health Organization) Regional Office for Africa. 2006. "Country Profiles on Environmental Health Policy." WHO Regional Office for Africa, Brazzaville. http://www.afro.who.int/des/phe/ country_profiles/index.html.

WHO and UNICEF (World Health Organization and United Nations Children's Fund). 2005. *Water for Life: Making It Happen.* Geneva: WHO and UNICEF Joint Monitoring Programme for Water Supply and Sanitation. http://www.wssinfo.org/en/40_wfl_2005.html.

Wierzba, T., R. Abu El-Yazeed, S. Savarino, A. Mourad, M. Rao, M. Baddour, A. N. El-Deen, A. Naficy, and J. Clemens. 2001. "The Interrelationship of Malnutrition and Diarrhea in a Periurban Area outside Alexandria, Egypt." *Journal of Pediatric Gastroenterology and Nutrition* 32 (2): 189–96.

Wilson, S. 1986. "Immunologic Basis for Increased Susceptibility of the Neonate to Infection." *Journal of Pediatrics* 108 (1): 1–9.

World Bank. 2000. "Eritrea Integrated Early Childhood Development Project." Report 20373-ER. Washington, DC, World Bank. http://www-wds.worldbank.org/external/default/WDSContentServer/WDSP/IB/2000/07/29/000094946_0006020538277/Rendered/PDF/multi_page.pdf.

————. 2002. "Arab Republic of Egypt: Cost Assessment of Environmental Degradation— Sector Note." Report 25175-EGT, World Bank, Washington, DC.

————. 2005. "Islamic Republic of Iran: Cost Assessment of Environmental Degradation— Sector Note." Report 32043-IR, World Bank, Washington, DC.

————. 2006a. *Environmental Priorities and Poverty Reduction: A Country Environmental Analysis for Colombia.* Washington, DC: World Bank.

————. 2006b. *Ghana: Country Environmental Analysis.* Washington, DC: World Bank.

————. 2006c. *Repositioning Nutrition as Central to Development: A Strategy for Large-Scale Action.* Washington, DC: World Bank.

————. 2007a. *Cost of Pollution in China: Economic Estimates of Physical Damages.* Washington, DC: World Bank.

————. 2007b. "Ghana—Nutrition and Malaria Control for Child Survival Project." Project appraisal document, World Bank, Washington, DC.

————. 2007c. *Health, Nutrition, and Population Sector Strategy Paper.* Washington, DC: World Bank.

————. 2007d. *Improving Traditional Hearths: Cleaner Air and Better Health in India.* Washington, DC: World Bank.

————. 2007e. *Key Issues in Central America Health Reforms: Diagnosis and Implications.* Report 36426-LAC, World Bank, Washington, DC.

————. 2007f. *Nutritional Failure in Ecuador: Causes, Consequences, and Solutions.* Washington, DC: World Bank.

————. 2007g. "Republic of Peru Environmental Sustainability: A Key to Poverty Reduction in Peru: Country Environmental Assessment." Report 40190-PE, World Bank, Washington, DC. http://go.worldbank.org/LDDPJN2TU0.

————. 2007h. *World Development Indicators.* Washington, DC: World Bank.

————. 2008. *Poverty and the Environment: Understanding Linkages at the Household Level.* Washington, DC: World Bank.

World Resources Institute, United Nations Environment Programme, United Nations Development Programme, and World Bank. 1998. *World Resources, 1998–99: A Guide to the Global Environment: Environmental Change and Human Health.* New York: Oxford University Press.

World Water Forum. 2006. "Water Infrastructures for Sustainable and Equitable Development." Framework Theme 1.34, Fourth World Water Forum, Mexico City, March 16–22. http://www.worldwatercouncil.org/index.php?id=1196.

WSP (Water and Sanitation Program). 2005. *Lessons Learned from Bangladesh, India, and Pakistan: Scaling-Up Rural Sanitation in South Asia.* New Delhi: WSP, World Bank.

————. 2006. *The Mumbai Slum Sanitation Program: Partnering with Slum Communities for Sustainable Sanitation in a Megalopolis.* New Delhi: WSP, World Bank.

————. 2007. *From Burden to Communal Responsibility: A Sanitation Success Story from Southern Region in Ethiopia.* New Delhi: WSP, World Bank.

Yacoob, M., and M. Kelly. 1999. "Secondary Cities in West Africa: The Challenge for Environmental Health and Prevention." Comparative Urban Studies Occasional Paper 21, Woodrow Wilson International Center for Scholars, Washington, DC.

Zaracostas, J. 2007. "Mortality from Measles Fell by 91% in Africa and by 68% Worldwide from 2000 to 2006." *British Medical Journal* 335: 1173.

Zhang, J., and K. Smith. 2005. "Indoor Air Pollution from Household Fuel Combustion in China: A Review." Presentation at Indoor Air 2005, Beijing, September 4–9.

Zumrawi, F., H. Dimond, and J. Waterlow. 1987. "Effects of Infection on Growth in Sudanese Children." *Human Nutrition and Clinical Nutrition* 41 (6): 453–61.

INDEX

Boxes, figures, notes, and tables are indicated by b, f, n, and t, respectively.

developing countries
  See also specific countries
  environmental health experiences in, 10–13,
    83–110
  health agendas in, 89–90, 101–4
  infectious disease studies in, 126*t*–140*t*
  infrastructure programs in, 94–96, 109, 113
  integrated child survival programs in,
    90–91, 109
  nutrition programs in, 91–94, 109, 113
  vector-control programs in, 96, 112
devolution of authority, 105–6
deworming, 22*b*–23*b*, 92, 118, 122–23
  See also helminth infections
DHS. *See* Demographic and Health Surveys
diarrhea
  antibiotics for treating, 35
  burden of disease, 51, 156
  cognition and, 29
  dehydration and, 20
  food restriction and, 122
  growth faltering and, 120–21, 124
  health systems and, 32, 33*t*
  household prevention of, 28
  in infancy, 22*b*, 27, 36, 117
  infections and, 17, 21, 123
  malnutrition and, 21*t*, 22*b*–23*b*, 68
  micronutrient supplementation and, 37, 38
  mortality and, 1, 55*f*, 55–56, 56*b*, 57, 59
  transmission routes, 18–19
  underweight and, 50*b*, 118, 125*n*4, 158, 158*t*
disability-adjusted life years (DALYs), 48,
    51–52, 53*f*, 148*t*–149*t*, 173
diseases. *See specific diseases by name*
drug resistance, 35, 37
dysentery, 21, 50*b*, 120, 124, 125*n*2

E

early infancy, 26–28, 27*f*, 30
economic costs, 9–10, 12
  cost-benefit analysis, 80–81, 114
  country-level costs of environmental health
    burden, 51–58
  environmental health valuations, 61–82
    See also environmental health valuations
Ecuador, nutrition programs in, 91, 94
education
  malnutrition and, 76*t*, 76–77, 81, 141–46,
    142*t*–143*t*, 160–70
  school performance, 29, 160, 162–64
  stunting and, 161*t*
  valuation of, 63–64, 76*t*, 76–77, 81, 112,
    141–46, 164–70

Egypt
  cost of environmental health risks in, 62, 63*f*
  nutritional status and infectious disease
    in, 138*t*
El Salvador, indoor air pollution in, 62
empowerment, 11, 98–99, 110, 113
energy and energy supplies. *See* fuels
environmental health, defined, 2*b*, 5–6
environmental health approaches, 85–110
  critical moments in, 109, 110
  decline in agenda for, 10, 88–90, 109, 113
  developing country experiences, 10–13,
    90–97, 109, 112–13
  enabling environment, 97–100
  governance and institutional implications,
    100–104
  history, 86–88
  institutional requirements, 104–9, 110, 113
environmental health valuations, 61–82,
    179–81
  cognition and, 63–64
  current practices, 61–62, 63*f*
  education and, 63–64, 76*t*, 76–77, 81, 112,
    141–46, 164–70
  malnutrition and, 9–10, 63, 66–78, 68*t*, 74*f*,
    74*t*, 76*t*, 81
  new estimates for, 62–64, 77–79, 79*t*, 80*b*,
    151–71, 154*t*, 179–81
  next steps, 79–81
environmental management programs, 9,
    38–39
epidemiology, 8–9, 12, 15–43
Eritrea Integrated Early Childhood
    Development Project, 91
Ethiopia
  environmental health institutions in, 102,
    102*b*–103*b*
  environmental health policies in, 104
  political commitment in, 98
  sanitation in, 12, 95, 95*b*, 102*b*
Europe
  developing countries compared to, 52,
    55*f*, 180
  malaria in, 6
Expanded Program on Immunization, 36

F

fecal pollution, 18, 19*f*, 28, 39
"fetal-programming," 25*b*, 58
  See also Barker hypothesis
fetuses, 7, 24–26, 25*b*, 58, 59
  See also pregnancy
Fewtrell, L., 51, 56*b*, 57, 112, 123
Fishman, S., 69, 176

child survival programs and, 90–91
diarrhea and, 38
environmental risk factors and, 61
infrastructure improvements and, 40, 41
maternal and childhood underweight, 50, 50b
miasma theory and, 86
pathogen transmission and, 28
private sector involvement in, 100
related diseases, 49t, 174t
water quantities and, 39–40

IAP. *See* indoor air pollution
IMCI. *See* Integrated Management of
    Childhood Illness
immunity, 20, 20f
immunizations, 36–37, 106, 112
    See also vaccinations
income losses, 68, 77, 165, 165t, 180, 181
India
    birth weight in, 54b
    environmental health functions in, 101b,
        101–2
    indoor air pollution in, 42
    Integrated Child Development Services, 91,
        93
    malaria in, 107
    malnutrition in, 6
    nutritional programs in, 91
    oral rehydration therapy in, 35
    sanitation in, 95, 99, 107, 110n1
Indonesia, 37, 134t
indoor air pollution (IAP)
    ALRI and, 1, 42, 50, 57, 75b, 174–77
    burden of disease, 49, 52
    clinical studies of, 58
    cookstoves and, 42–43, 95–96
    environmental risk factors of, 6, 61
    malaria and, 39
    mortality, 59n7
    pregnancy and, 26
    related diseases, 49t, 174t
indoor residual spraying, 28, 38, 57, 96, 113
infant care programs, 35–36
infants and young children
    See also specific diseases and interventions
    breastfeeding. *See* breastfeeding
    burden of disease, 48–49, 53f
    cognition and, 28–29
    diarrhea and, 22b, 27, 36, 117
    environmental health risks and, 1, 2b, 8,
        26–28
    infections in, 20–21

linear growth in, 23, 25b
malnutrition and, 19, 26–28, 27f, 30, 68t,
    68–76
infections
    diarrhea and, 17, 21, 123
    in early infancy, 26–28, 35–36
    growth faltering and, 117, 119–20
    malnutrition and, 19–23, 20f, 30, 51, 63,
        118, 123–24, 126t–140t
    micronutrient supplementation and, 37–38
    during pregnancy, 24–26, 25b
    primary prevention of, 43
infrastructure
    child survival strategies and, 9, 39–43
    developing countries' programs, 94–96, 109,
        113
    public health agenda and, 89
insecticide-treated nets (ITNs), 38–39,
    96, 113
institutional capacity, 80
Integrated Child Development Services
    (India), 91, 93
Integrated Community-Based Child Care
    Program (Honduras), 91
Integrated Early Childhood Development
    Project (Eritrea), 91
Integrated Management of Childhood Illness
    (IMCI), 32, 90–91
intersectoral and interjurisdictional
    coordination, 13, 104–5, 107, 108–9,
    114–15
interventions
    burdens of disease and, 48
    childhood growth and, 23
    costs and benefits of, 80
    customizing of, 11
    by health ministries, 89
    in history of environmental health, 86–87
    multiplier effect for environmental health,
        55–56
intrauterine growth restriction (IUGR), 7
invisible regulator, 87–88
ITNs. *See* insecticide-treated nets
IUGR (intrauterine growth restriction), 7

Jacoby, H., 145
Jamaica
    correlation between education deficiency
        and stunting, 144
    helminth infections in, 29
Jonsson, U., 54b
jurisdictional collaborations. *See* intersectoral
    and interjurisdictional coordination

ECO-AUDIT
Environmental Benefits Statement

The World Bank is committed to preserving endangered forests and natural resources. The Office of the Publisher has chosen to print *Environmental Health and Child Survival: Epidemiology, Economics, Experiences* on recycled paper with 100 percent postconsumer fiber in accordance with the recommended standards for paper usage set by the Green Press Initiative, a nonprofit program supporting publishers in using fiber that is not sourced from endangered forests. For more information, visit www.greenpressinitiative.org.

Saved:
- 48 trees
- 33 million BTUs of total energy
- 4,216 lbs. of CO_2 equivalent of greenhouse gases
- 20,650 gallons of waste water
- 2,247 pounds of solid waste